Lecture Notes in Economics and Mathematical Systems

Managing Editors: M. Beckmann and W. Krelle

252

Alessandro Birolini

On the Use of Stochastic Processes in Modeling Reliability Problems

Springer-Verlag
Berlin Heidelberg New York Tokyo

Author

Dr. Alessandro Birolini
PD at the ETH Zurich and
Product Assurance Consultant
CH-8606 Nänikon, Switzerland

ISBN 3-540-15699-2 Springer-Verlag Berlin Heidelberg New York Tokyo
ISBN 0-387-15699-2 Springer-Verlag New York Heidelberg Berlin Tokyo

Printing and binding: Beltz Offsetdruck, Hemsbach/Bergstr.
2142/3140-543210

PREFACE

Stochastic processes are powerful tools for the investigation of reliability and availability of repairable equipment and systems. Because of the involved models, and in order to be mathematically tractable, these processes are generally confined to the class of regenerative stochastic processes with a finite state space, to which belong: renewal processes, Markov processes, semi-Markov processes, and more general regenerative processes with only one (or a few) regeneration state(s). The object of this monograph is to review these processes and to use them in solving some reliability problems encountered in practical applications. Emphasis is given to a comprehensive exposition of the analytical procedures, to the limitations involved, and to the unification and extension of the models known in the literature. The models investigated here assume that systems have only one repair crew and that no further failure can occur at system down. Repair and failure rates are generalized step-by-step, up to the case in which the involved process is regenerative with only one (or a few) regeneration state(s). Investigations deal with different kinds of reliabilities and availabilities for series/parallel structures. Preventive maintenance and imperfect switching are considered in some examples.

This monograph is based upon subject material from a course which I have been teaching since 1975 at the department of electrical engineering of the Swiss Federal Institute of Technology (ETH) Zurich. The text is intended for research workers or theoretically oriented engineers in industry, as well as for graduate students in electrical engineering, industrial or mechanical engineering, operations research, computer science, mathematics, and economics. For the non-specialist, chapter 2 contains a brief introduction to the basic concepts associated with reliability analysis, as viewed from an engineering standpoint. Greater detail on this subject can be found in [9].

This work also stands as a habilitation thesis for the department of electrical engineering at the ETH Zurich and has been accepted there, in a preliminary version, in December, 1984. I would like to express my gratitude to Prof. Dr. M. Mansour (main referee) and to Prof. Dr. J. Hugel (referee) for supporting this thesis, as well as to Dr. U. Höfle-Isphording, Prof. Dr. J. Kohlas, and the Reviewer of Springer-Verlag for their advice and criticism. My thanks are also due to Springer-Verlag for the very pleasant cooperation.

Nänikon near Zurich
Spring, 1985

A. Birolini

CONTENTS

CHAPTER 1

INTRODUCTION AND SUMMARY

For complex equipment and systems, reliability analysis is generally performed at two differents levels. At subassembly level, the designer performs failure rate and failure mode analyses to check fulfilment of reliability requirements, and to detect and eliminate reliability weaknesses as early as possible in the design phase. At equipment and system level, the reliability engineer also investigates time behaviour, taking into account reliability, maintainability, and logistical aspects. Depending upon the system complexity, upon the assumed distribution functions for failure-free and repair times, and with thought toward maintenance policy, investigations are performed either analytically, making use of stochastic processes, or numerically with the help of Monte Carlo simulations. Stochastic processes used in the modeling of reliability problems include renewal and alternating renewal processes, Markov processes with a finite state space, semi-Markov processes, regenerative stochastic processes with only one (or a few) regeneration state(s), and some kinds of non-regenerative stochastic processes. The reliability models covered by each of these processes are given in Table 1.

Stochastic processes	Can be used in modeling	Background	Degree of difficulty
Renewal processes	Spare parts reservation in the case of arbitrary failure rates and negligible replacement or repair times	Renewal theory	Medium
Alternating renewal processes	One-item renewable structures with arbitrary failure and repair rates	Renewal theory	Medium
Markov processes (finite state space, time-homogeneous)	Systems of arbitrary structure whose elements have constant failure and repair rates	Differential equations	Low
Semi-Markov processes (at least embedded)	Some systems whose elements have constant failure rates and arbitrary repair rates	Integral equations	Medium
Regenerative processes with only one (or a few) regeneration state(s)	Systems of arbitrary structure whose elements have constant failure rates and arbitrary repair rates; some reduntant structures whose elements have quite general failure rates and arbitrary repair rates	Integral equations	High
Non regenerative processes (e.g. superimposed renewal or alternating renewal processes)	Systems of arbitrary structure whose elements have arbitrary failure and repair rates	Sophisticated methods, partial differential equations	High to very high

Table 1. Stochastic processes used in reliability and availability analysis

After chapter 2's brief review of the basic concepts associated with reliability analysis, chapter 3 will introduce the stochastic processes given in Table 1, and chapters 4 through 6 will then use them in solving some reliability problems encountered in practical applications. Emphasis is given to a comprehensive exposition of the analytical procedures, to the theoretical and computational limitations involved, and to the unification and extension of the models known in the literature.

The one-item repairable structure is investigated in chapter 4 by assuming arbitrary distribution functions for the failure-free and repair times, as well as arbitrary initial conditions at $t = 0$. These assumptions allow a careful analysis of the asymptotic and stationary behaviours. The investigation deals with different kinds of reliabilities and availabilities. Series, parallel and series/parallel structures are investigated in chapter 5. For these models it is assumed that systems have only one repair crew and that no further failure can occur at system down. Failure-free and repair time distribution functions are generalized step-by-step, starting with the exponential distribution, up to the case in which the involved process is regenerative with only one (or a few) regeneration state(s). Expressions are derived for the reliability function, the mean time-to-system-failure, the point-availability, and, as far as possible, the interval-reliability. Asymptotic and stationary behaviours are discussed for each model. The influence of the repair times density shape on the mean time-to-system-failure and on the stationary value of point-availability is investigated. Methods applicable to more complex structures are outlined in section 6.1. The effects of preventive maintenance and imperfect switching on reliability and availability are considered for some examples in sections 6.2 and 6.3.

The main results obtained in this study can be summarized as follows:

1. Stochastic processes are powerful tools for the investigation of reliability problems - in particular, for investigation of the time behaviour of repairable equipment and systems. If all elements in the system are independent and have constant failure and repair rates, time-homogeneous Markov processes with a finite state space can be used. Markov processes also arise if the repair and/or failure-free times have an Erlang distribution (supplementary states). If repair rates are general, but the system has only one repair crew, the involved process contains an embedded semi-Markov process. A further generalisation leads to processes with only one (or a few) regeneration state(s), or even to non-regenerative stochastic processes.

2. The method of investigation applicable to stochastic processes with only one (or a few) regeneration state(s) is based on the fact that, between consecutive occurrence points of the same regeneration state (cycle), the development of the process is a probabilistic replica of the development between the first and second of these regeneration points. Thus, the analysis can be limited to only one cycle

by considering two cases: (1) no system failure occurs and (2) system failure occurs. Case (1) leads to the reliability function, cases (1) and (2) lead to point-availability. The time interval between $t=0$ and the first occurrence point of the regeneration state has to be considered only if the system does not enter the regeneration state at $t=0$. If the stochastic process is not regenerative, supplementary variables can be used. However, such an approach generally involves complicated partial differential equations.

3. Besides the analytical limitations outlined in point 2, computational difficulties arise for large systems even if Markov or semi-Markov processes can be used. Therefore, great importance must also be paid in the future to the investigation of practice-oriented models, to the assumption of stationary states, and to the use of approximation expressions obtained by series expansions or as limit expressions. The procedures and models investigated in chapters 3 through 6 consider these aspects. As shown, the reliability investigations are simplified by considering Markov processes as particular cases of semi-Markov processes. Generalization of the initial conditions at $t=0$ allows investigation and comparison of the asymptotic behaviour and the stationary state. The assumption of no further occurring failures at system down simplifies investigations of availability and interval-reliability. The influence of the repair times density shape on mean time-to-system-failure and on the stationary value of point-availability is small if, for each element in the system, the mean time-to-repair is much shorter than the mean time-to-failure (MTTR $\ll$ MTTF).

CHAPTER 2

BASIC CONCEPTS OF RELIABILITY ANALYSIS

This chapter introduces the basic concepts associated with reliability analysis, taking as an example the investigation of non-repairable systems. An important part of such an investigation deals with failure rate and failure mode analyses. The failure mode analysis will be discussed in section 2.5. The failure rate analysis leads to the predicted reliability, i.e. to that reliability estimate obtained analytically from the reliability structure of the system and the failure rate of its elements. The prediction is useful to detect reliability weaknesses, to quantify the utility of reliability improvements (derating, screening, redundancy) and to compare alternative solutions. It is performed according to the following procedure:

1. Definition of the system and of its associated mission profile; derivation of the corresponding reliability block diagram.
2. Determination of the stresses applied and of the corresponding failure rate (λ) for each element of the reliability block diagram.
3. Evaluation of the system reliability function ($R_S(t)$).
4. Elimination of reliability weaknesses and return to step 1 or 2, as long as necessary.

Steps 1 to 3 are discussed in the following sections.

2.1 Mission profile, reliability block diagram

The mission profile specifies the task to be accomplished (required function), its variation with time, and the associated environmental conditions. To each mission profile corresponds a reliability block diagram. This diagram gives the answer to the following question: Which elements of the system are necessary for the required function and which ones can fail without affecting it (redundancy)? The necessary elements are put in series and redundancies appear in parallel on the reliability block diagram. In setting up the reliability block diagram, care must be taken regarding the fact that, for each element, only two states (good or failed) and one failure mode (e.g. open or short) can be considered.

As an example, Fig.1 gives functional and reliability block diagrams of an electronic switch for two different situations: case a) without redundancy and case b) with redundancy, on the transistor. In case b), the assumed failure mode of the transistors is a short between collector and emitter (the failure mode of resistors is open).

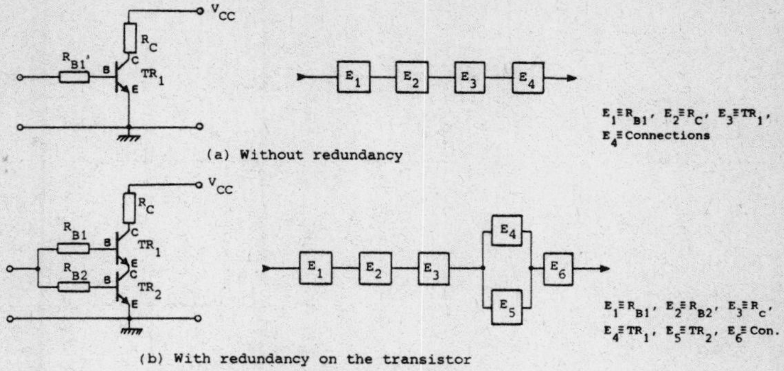

Fig. 1. **Functional and reliability block diagrams for an electronic switch (assumed failure mode for the transistors: short between C and E)**

For large equipment or systems, the reliability block diagram is derived top down as indicated in Fig.2.

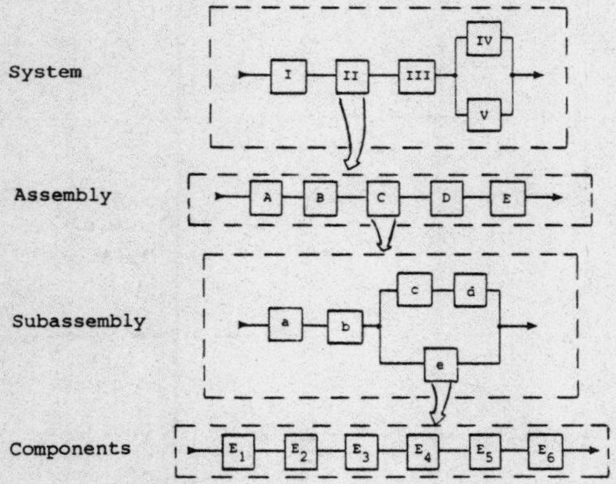

Fig. 2. **Procedure for setting up the reliability block diagram of a system**

The typical structures of reliability block diagrams are summarized in Table 2 on page 6. Also given in Table 2 are the associated reliability functions for the case of non-repairable items (systems) with active redundancy and statistically independent elements (see section 2.3 for derivation).

2.2 Failure rate

To introduce the mathematical concept of failure rate, let τ be the failure-free time of a given item and $F(t)$ its distribution function with density $f(t) = dF(t)/dt$. (Failure-free and repair times will be considered in this monograph as positive and continuous random variables.) For this item, the reliability function is given by

Reliability block diagramm	Reliability function ($R_S = R_S(t)$, $R_i = R_i(t)$)	Remarks
	$R_S = R_i$	One-item structure
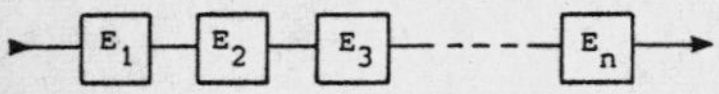	$R_S = \prod_{i=1}^{n} R_i$	Series structure
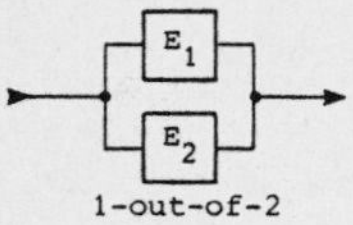	$R_S = R_1+R_2-R_1R_2$	1-out-of-2 redundancy
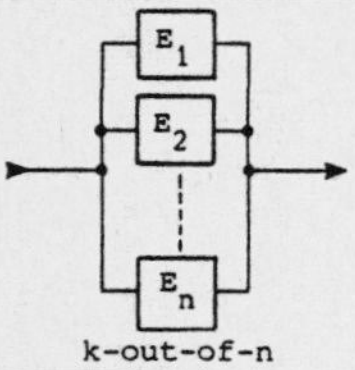	$R_1 = R_2 = \ldots = R_n = R$ $R_S = \sum_{i=k}^{n} \binom{n}{i} R^i (1-R)^{n-i}$	k-out-of-n redundancy
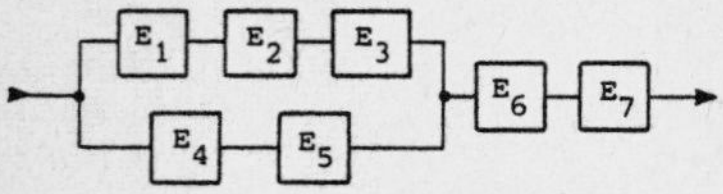	$R_S = (R_1R_2R_3+R_4R_5 -R_1R_2R_3R_4R_5)R_6R_7$	Series/parallelstructure
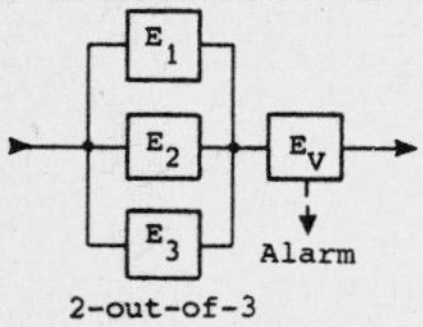	$R_1 = R_2 = R_3 = R$ $R_S = (3R^2-2R^3)R_V$	Majority redundancy (general case: (n+1)-out-of-(2n+1))
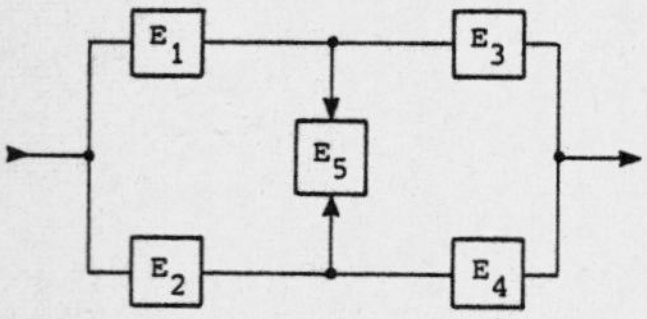	$R_S = R_5(R_1+R_2-R_1R_2)(R_3+R_4-R_3R_4) +(1-R_5)(R_1R_3+R_2R_4-R_1R_2R_3R_4)$	Bridge structure (bidirectional on E_5)
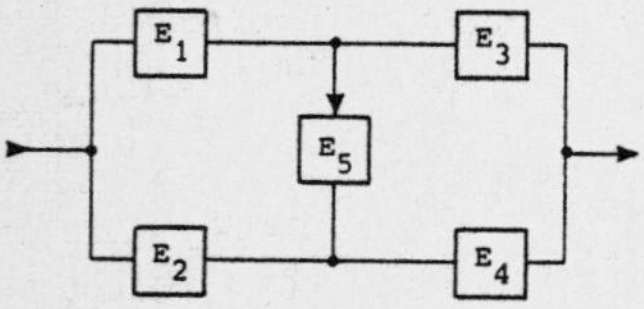	$R_S = R_4[R_2+R_1(R_3+R_5-R_3R_5)-R_1R_2 (R_3+R_5-R_3R_5)]+(1-R_4)R_1R_3$	Bridge structure (unidirectional on E_5)
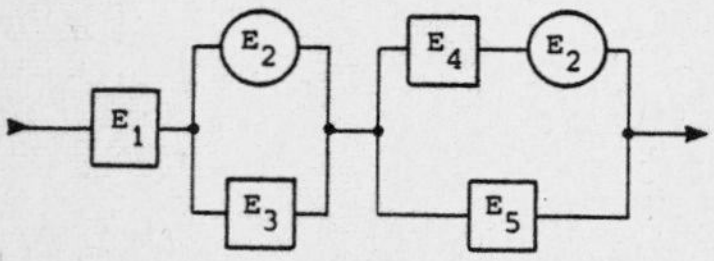	$R_S = R_2R_1(R_4+R_5-R_4R_5) +(1-R_2)R_1R_3R_5$	Element E_2 appear twice in the reliability block diagram

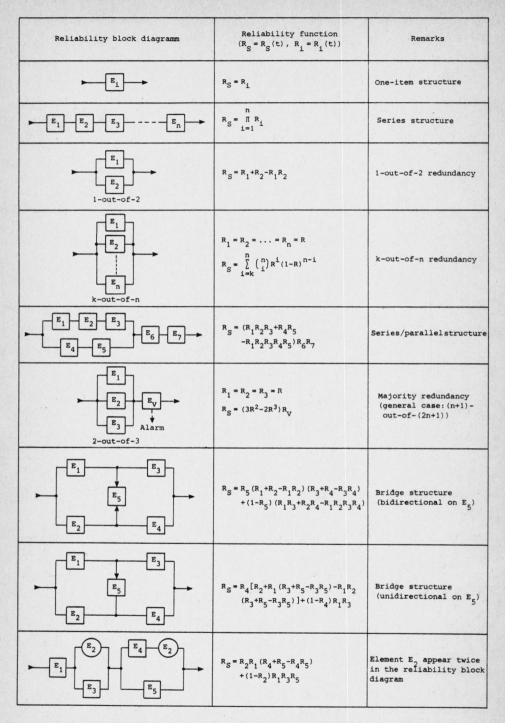

Table 2. Typical reliability block diagram structures and associated reliability functions (non-repairable systems, active redundancy, independent elements)

$$R(t) = \Pr\{\tau > t\} = 1-F(t). \qquad (1)$$

Assume now that the item is put in the operating state at $t=0$, and that it works without failure up to the time t. The probability that it will fail in the next time interval δt is given by

$$\Pr\{t < \tau \leq t+\delta t \mid \tau > t\} = \frac{\Pr\{t < \tau \leq t+\delta t \cap \tau > t\}}{\Pr\{\tau > t\}} = \frac{\Pr\{t < \tau \leq t+\delta t\}}{\Pr\{\tau > t\}}. \qquad (2)$$

The limit for $\delta t \to 0$ of this probability divided by δt is the failure rate $\lambda(t)$

$$\lambda(t) = \lim_{\delta t \to 0} \frac{1}{\delta t} \Pr\{t < \tau \leq t+\delta t \mid \tau > t\}, \qquad \delta t > 0. \qquad (3)$$

From equations (1), (2) and (3) it follows that

$$\lambda(t) = \frac{f(t)}{1-F(t)} = -\frac{dR(t)/dt}{R(t)} \qquad (4)$$

and, with $R(0) = 1$,

$$R(t) = e^{-\int_0^t \lambda(x)dx}. \qquad (5)$$

The failure rate of a population of statistically identical items is generally characterized by a period of early failures, a period with approximately constant failure rate, and a wear-out period (bath-tub curve). For an item with constant failure rate

$$\lambda(t) = \lambda, \qquad t \geq 0, \qquad (6)$$

equation (5) leads to

$$R(t) = e^{-\lambda t}. \qquad (7)$$

In this case, the quantity $\lambda(t)\delta t$, i.e.

$$\lim_{\delta t \to 0} \Pr\{\text{failure in interval } (t,t+\delta t) \mid \text{no failure up to the time } t\},$$

is equal to $\lambda \delta t$ independent of t. This *memoryless* property is a characteristic feature of the exponential distribution function $F(t) = 1-e^{-\lambda t}$, and will be used in the following chapters to check the presence of *regeneration points*.

The failure rate of a given item depends on the electrical, thermal, climatic, and mechanical stresses applied to it, as well as on the technology and on the production quality. For electronic components it lies between $10^{-9}h^{-1}$ and $10^{-6}h^{-1}$.

Corresponding models have been worked out [11,40]. These models assume a constant failure rate (λ) and are given by

$$\lambda = \lambda_b \pi_E \pi_Q \pi_A \tag{8}$$

for discrete components and

$$\lambda = [C_1 \pi_T \pi_V \pi_{PT} + (C_2 + C_3)\pi_E]\pi_Q \pi_L \tag{9}$$

for integrated circuits (ICs).

In equation (8), λ_b is the basic failure rate. It depends on the ambient temperature (θ_A) and on the stress factor (S). Fig.3 gives examples for some discrete components. For ICs, the corresponding factors are, according to equation (9): C_1 and C_2 for the complexity (Fig.4), C_3 for the package size and type (Fig.5), π_T for the chip temperature (Fig.6), and π_V for the voltage stress (Fig.7). π_V is only defined for CMOS-ICs. π_T can be approximately evaluated by using the Arrhenius model

$$\frac{\pi_{T_2}}{\pi_{T_1}} = e^{\frac{E_A}{k}\left(\frac{1}{T_1} - \frac{1}{T_2}\right)}, \tag{10}$$

with activation energy (E_A) between 0.4 and 0.9eV and $k = 8.6 \cdot 10^{-5}$eV/K.

The factor π_E takes care of environmental conditions (Shock, vibration, moisture, pressure, noise, etc.). As shown in Table 3 it is also dependent upon the technology.

Environment	Factor π_E				
	ICs Monol.	ICs Hybr.	Discrete semiconductors	Resistors	Capacitors
G_B (Ground Begnin)	0.38	0.2	1	1	1
G_F (Ground Fixed)	2.5	0.78	2.4 - 5.8	1.5 - 2.9	1.4 - 2.4
G_M (Ground Mobile)	4.2	2.2	7.8 - 18	7.8 - 11	7.8 - 12

***Table 3.* Factor π_E for some important environmental conditions (MIL-HDBK-217 [40])**

The factor π_Q is a function of production quality and of the adopted screening procedure. It is given in Table 4 for some important electronic components. A typical screening procedure for ICs in ceramic packages according to class B2 of Table 4 consists of: 24 hours high temperature storage at 150°C, 10 thermal cycles -65/+150°C,

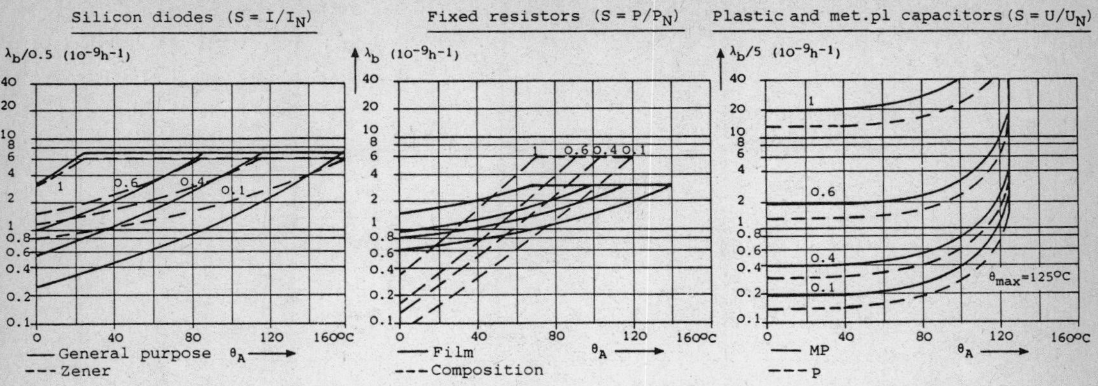

Fig. 3. Basic failure rate (λ_b) for some discrete electronic components as a function of the ambient temperature (θ_A), and with the stress factor S = 0.1, 0.4, 0.6, and 1 as a parameter (MIL-HDBK-217[40])

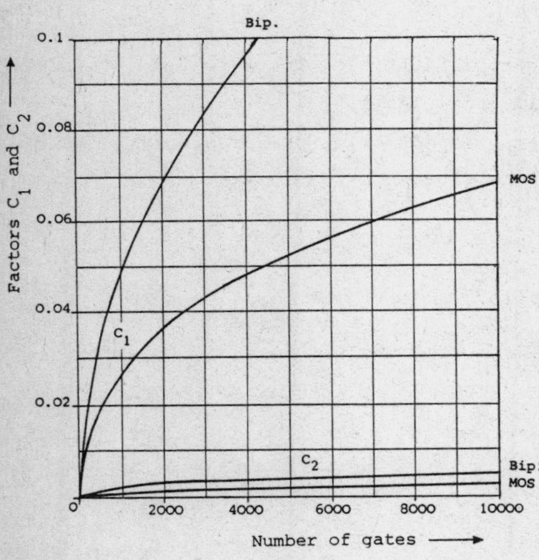

Fig. 4. Factors C_1 and C_2 for ICs (MIL-HDBK-217[40])

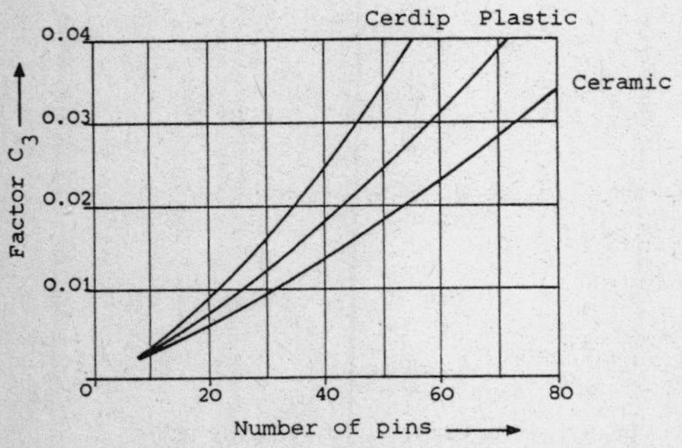

Fig. 5. Factor C_3 for ICs (MIL-HDBK-217[40])

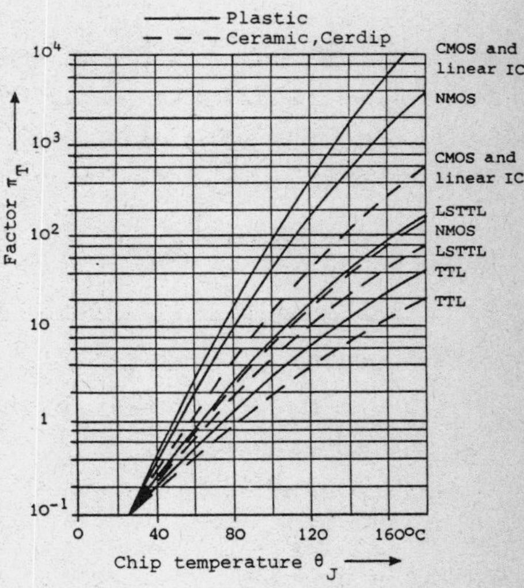

Fig. 6. Factor π_T for ICs (MIL-HDBK-217[40])

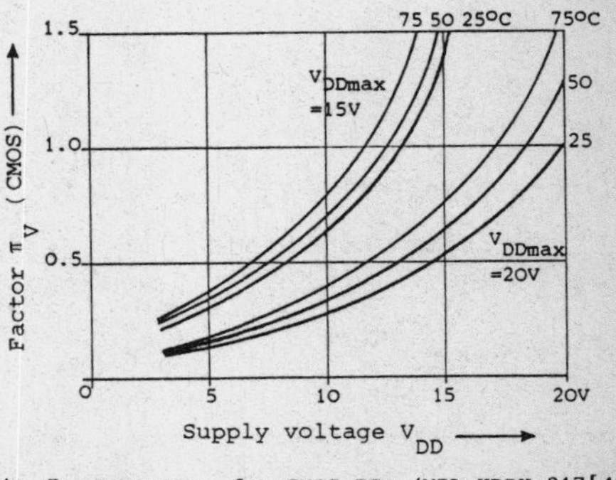

Fig. 7. Factor π_V for CMOS-ICs (MIL-HDBK-217[40])

Component	Class Factor π_Q								
Monolithic ICs	S 0.5	B 1	B-O 2	B-1 3	B-2 6.5	C 8	C-1 13	D 17.5	D-1 35
Hybrid ICs	B 1								
Discrete semi-conductors	JANTXV 0.12	JANTX 0.24	JAN 1.2	Lower 6	Plastic 12				
Resistors	S 0.03	R 0.1	P 0.3	M 1	NE 5	Lower 15			
Capacitors	S 0.03	R 0.1	P 0.3	M 1	L 1.5	NE 3	Lower 10		

Table 4. **Factor π_Q for some important electronic components (MIL-HDBK-217 [40])**

60 sec constant acceleration at 30'000g, electrical test, 160 hours burn-in at 125°C, final electrical test, and seal test (fine and gross leak).

Finally, from equation (9), π_{PT} and π_L are, respectively, the programming technique and the learning factor ($1 \leq \pi_{PT} \leq 5$, $1 \leq \pi_L \leq 10$).

2.3 Reliability function, MTTF, MTBF

The reliability function R(t) gives, for each value of the parameter t, the probability that no failure occurs during the time interval (0,t). For a one-item structure (new at t = 0), equation (1) holds. For such an item, the mean time-to-failure (MTTF or simply MTF) is given by

$$MTTF = E\{\tau\} = \int_0^\infty R(t)\,dt, \tag{11}$$

if its useful life is taken to be ∞, and

$$MTTF_L = \int_0^{T_L} R(t)\,dt, \tag{12}$$

if its useful life is limited to T_L. In the following it will be assumed $T_L = \infty$.

A particular situation arises when

$$R(t) = e^{-\lambda t}$$

holds (constant failure rate $\lambda(t) = \lambda$). In this case, equation (11) leads to

$$E\{\tau\} = 1/\lambda.$$

As in the literature, we shall set

$$\frac{1}{\lambda} = \text{MTBF}, \tag{13}$$

where MTBF stands for mean time-between-failures (for $\lambda(t) = \lambda$).

To evaluate the reliability function $R_S(t)$ of the models given in Table 2 (p.6) let us assume that each element of the reliability block diagram is non-repairable and works independently of any other element (non-repairable systems with active redundancy and independent elements).

For the series structure, the system has no failures in the interval (0,t) only if each element has no failures in this interval. This leads to

$$\begin{aligned} R_S(t) &= \Pr\{\text{system up in } (0,t)\} \\ &= \Pr\{E_1 \text{ up in } (0,t) \cap E_2 \text{ up in } (0,t) \cap \ldots \cap E_n \text{ up in } (0,t)\}. \end{aligned} \tag{14}$$

Because of the assumed independence, it follows from equation (14)

$$R_S(t) = R_1(t)R_2(t)R_3(t)\ldots R_n(t) = \prod_{i=1}^{n} R_i(t). \tag{15}$$

Using the concept of failure rate (eq.(5)), equation (15) becomes

$$R_S(t) = e^{-\int_0^t \lambda_S(x)\,dx}$$

with

$$\lambda_S(x) = \sum_{i=1}^{n} \lambda_i(x). \tag{16}$$

Relation (16) is useful for evaluation of the predicted reliability. It states that the failure rate of a non-redundant system which consists of independent elements is the sum of the failure rates of its elements.

In a similar manner, one obtains, for the case of a 1-out-of-2 active redundancy with independent elements,

$$\begin{aligned} R_S(t) &= \Pr\{E_1 \text{ up in } (0,t) \cup E_2 \text{ up in } (0,t)\} \\ &= R_1(t)+R_2(t)-R_1(t)R_2(t). \end{aligned} \tag{17}$$

The reliability function of a k-out-of-n active redundancy with n identical and independent elements can be considered as the probability of at least k successes in n Bernoulli trials with $p = R(t)$.

Series/parallel structures are investigated using, alternatively, equations (15) and (17).

The last three models of Table 2 cannot be reduced to series/parallel structures with independent elements. However, an easy way to compute their reliability functions is to use the theorem of total probability in the following form:

$$\begin{aligned} R_S(t) &= \Pr\{\text{system up in } (0,t)\} = \Pr\{E_i \text{ up in } (0,t) \cap \text{system up in } (0,t)\} \\ &\quad + \Pr\{E_i \text{ failed in } (0,t) \cap \text{system up in } (0,t)\} \\ &= R_i(t)\Pr\{\text{system up in } (0,t) \mid E_i \text{ up in } (0,t)\} \\ &\quad + (1-R_i(t))\Pr\{\text{system up in } (0,t) \mid E_i \text{ failed in } (0,t)\}. \end{aligned} \tag{18}$$

The element E_i must be chosen in such a way that a series/parallel structure is obtained for the reliability block diagrams conditioned upon the events *E_i up* and *E_i failed*. Successive application of equation (18) in a tree diagram is also possible, see equation (294). Other investigative methods for complex structures are based on Boolean representations [3,8,9,21,29,34,48,49,62].

2.4 More general considerations on the concept of redundancy

In the redundancy structures investigated in section 2.3, all elements are still operating at the same conditions. For this type of redundancy, called active (parallel or hot) redundancy, the assumed statistical independence of the elements implies in particular that there is no *load sharing*. This assumption does not arise in many practical applications, e.g. at component level or in the presence of power elements. The investigation of the reliability function in the case of load sharing or of other kinds of dependency involves the use of stochastic processes. The situation is simple if one can assume that the failure rate of each element changes only when a failure occurs. In this case, the general model for a k-out-of-n redundancy is given in Fig.8.

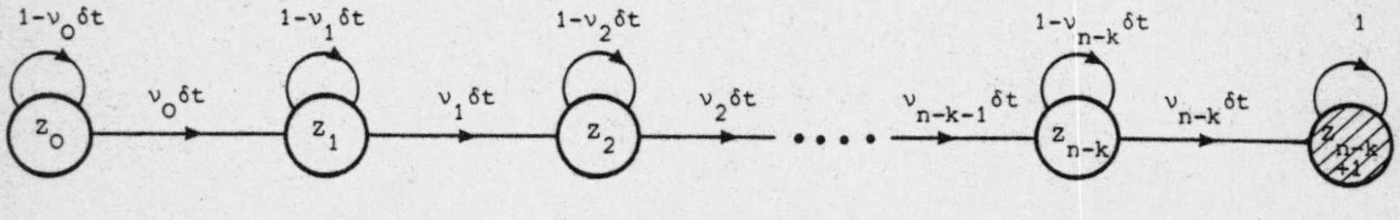

Fig. 8. **Diagram of the transition probabilities in (t,t+δt) for a non-repairable k-out-of-n redundancy (constant failure rate between successive state changes, t arbitrary, δt→0)**

$Z_0, Z_1, Z_2, Z_3, \ldots, Z_{n-k+1}$ are the states of the process. In state Z_i exactly i elements are down. At state Z_{n-k+1} the system is down. Assuming

λ = failure rate of an element in the operating state

and

λ_r = failure rate of an element in the reserve state ($\lambda_r < \lambda$),

the model of Fig.8 considers, in particular, the following situations:

1. Active redundancy without load sharing

$$\nu_i = (n-i)\lambda, \qquad i = 0,1,2,\ldots,n-k \tag{19}$$

λ is the same for all states.

2. Active redundancy with load sharing

$$\nu_i = (n-i)\lambda, \qquad i = 0,1,2,\ldots,n-k \tag{20}$$

λ increases at each state change.

3. Warm (lightly loaded) redundancy

$$\nu_i = k\lambda+(n-k-i)\lambda_r, \qquad i = 0,1,2,\ldots,n-k \tag{21}$$

λ and λ_r are the same for all states.

4. Standby (cold) redundancy

$$\nu_i = k\lambda, \qquad i = 0,1,2,\ldots,n-k \tag{22}$$

λ is the same for all states.

For a standby redundancy one assumes that the failure rate in the reserve state is $\equiv 0$ (the reserve elements are switched on when needed). Warm redundancy is between active and standby ($0 < \lambda_r < \lambda$). It should be noted that the k-out-of-n active, warm, or standby redundancy is only the simplest representative of the general concept of redundancy. Series/parallel structures, voting techniques, bridges, and more complex structures are frequently used. Redundancy can also appear in the coding, at software level, or in many other forms. Furthermore, the benefit of a redundancy can be limited by the involved failure modes, as well as by switching effects. For the investigation of the model of Fig.8, let

$$P_i(t) = \Pr\{\text{the process is in state } Z_i \text{ at time } t\}, \tag{23}$$

be the state probabilities ($i = 0,1,\ldots,n-k+1$). The functions $P_i(t)$ satisfy the fol-following system of differential equations:

$$\begin{aligned}
\dot{P}_0(t) &= -\nu_0 P_0(t) \\
\dot{P}_i(t) &= -\nu_i P_i(t)+\nu_{i-1}P_{i-1}(t), \qquad i = 1,2,3,\ldots,n-k \\
\dot{P}_{n-k+1}(t) &= \nu_{n-k}P_{n-k}(t).
\end{aligned} \tag{24}$$

$\dot{P}_i(t)$ is obtained by considering the process at two adjacent time points t and $t+\delta t$, and making use of the *memoryless* property given by the constant failure rate assumed between consecutive state changes. For $\delta t \to 0$ one obtains from Fig.8

$$P_i(t+\delta t) = P_i(t)(1-\nu_i\delta t)+P_{i-1}(t)\nu_{i-1}\delta t+o(\delta t), \tag{25}$$

which leads to equation (24). Knowing $P_i(t)$, one can evaluate the reliability function of the system from (see Fig.8)

$$R_S(t) = \sum_{i=0}^{n-k} P_i(t) = 1-P_{n-k+1}(t), \tag{26}$$

and the mean time-to-system-failure from

$$MTTF_S = \int_0^\infty R_S(t)dt. \tag{27}$$

Assuming $P_0(0)=1$ as an initial condition, one obtains for the Laplace transform of $R_S(t)$

$$\tilde{R}_S(s) = \int_0^\infty R_S(t)e^{-st}dt \tag{28}$$

the expression

$$\tilde{R}_S(s) = \frac{(s+\nu_0)(s+\nu_1)(s+\nu_2)\dots(s+\nu_{n-k})-\nu_0\nu_1\nu_2\dots\nu_{n-k}}{s(s+\nu_0)(s+\nu_1)(s+\nu_2)\dots(s+\nu_{n-k})}, \tag{29}$$

and for the mean time-to-system-failure

$$MTTF_S = \tilde{R}_S(0) \tag{30}$$

the expression

$$MTTF_S = \sum_{i=0}^{n-k} \frac{1}{\nu_i}. \tag{31}$$

The case of the standby redundancy, with ν_i given by equation (22), leads to

$$R_S(t) = \sum_{i=0}^{n-k} \frac{(k\lambda t)^i}{i!} e^{-k\lambda t}, \tag{32}$$

and

$$MTTF_S = \frac{n-k+1}{k\lambda}. \tag{33}$$

Equation (32) shows the relation existing between the Poisson distribution and the occurrence of exponentially distributed events.

For the case of active redundancy without load sharing, it follows from equations (31) and (19) that

$$MTTF_S = \frac{1}{\lambda}\left(\frac{1}{n}+\frac{1}{n-1}+\dots+\frac{1}{k}\right). \tag{34}$$

Expressions for $R_S(t)$ are given in Fig.9 for different values of n and k.

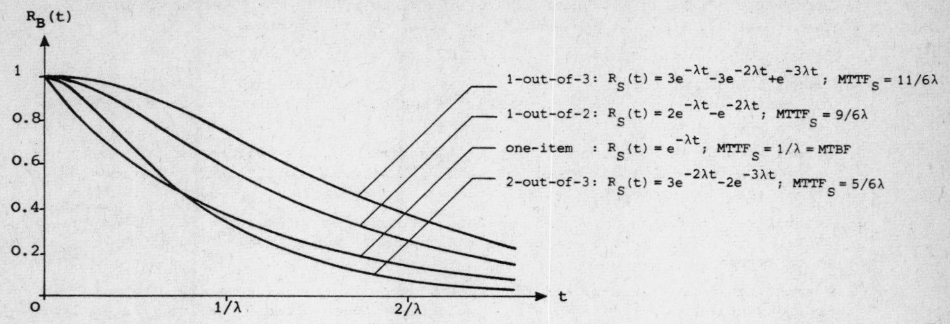

Fig. 9. **Reliability functions for the one-item structure and for some kinds of non-repairable k-out-of-n active redundancies (constant failure rates, identical elements, no load sharing)**

2.5 Failure mode analysis and other reliability assurance tasks

Failure rate analyses, as discussed in sections 2.1 to 2.4, do not account for the mode and the effect (consequence) of a failure. To better understand the mechanism of system failures and in order to identify the potential weaknesses, it is necessary for critical system parts to perform a *failure modes, effects and criticality analysis* (FMECA). The FMECA consists of a systematic analysis of all possible modes of failure, their causes, effects, and criticalities [1,8,9,41,42,52,66]. It also investigates means for avoiding the failures or for minimizing their consequences. It is performed bottom-up by the reliability engineer together with the designer. The procedure is easy to understand but time-consuming to apply, see Table 5 on p.16.

A further possibility for investigation of the failure-cause-to-effect-relations is the *fault trees analysis* (FTA). The FTA is a top-down procedure in which the undesired event, e.g. a critical or catastrophic failure at system level, is represented by AND and OR combinations of causes at lower levels [3,9,30,52,66]. The FTA can also consider external influences (human and/or environmental), and handle situations where more than one primary failure (or fault) has to occur to cause the undesired event at system level. However, it does not necessarily go through all possible failure modes. The combination of a fault trees analysis with a failure modes, effects and criticality analysis leads to a *cause-consequence chart* [66]. Such a chart shows the logical relationship between all identified causes and their single or multiple consequences.

The reliability analyses described in this chapter are an important part of the reliability assurance tasks during the design phase. To obtain maximum benefit, they must be complemented by other reliability assurance activities in the development and production phases. To these activities belong, in particular, the selection and qualification of components and parts as well as of production processes and procedures,

1. Serial numbering of the step.
2. Definition of the item or part in question (reference to the reliability block diagram, parts list etc.), and short description of its function.
3. Assumption of a failure mode (all possible failure modes must be considered).
4. Identification of possible causes for the failure mode assumed in point 3 (a cause for a failure can also be an error in the design or production phase: overload, material problem etc.).
5. Description of the symptom(s) which will characterize the failure mode assumed in point 3, and of the local effect of this failure mode (output/input relations, possibilities for secondary failures etc.).
6. Identification of the consequences of the failure mode assumed in point 3 on the next higher integration levels (up to the equipment or system level), and on the mission to be performed.
7. Identification of corrective actions which can mitigate the severity of the failure mode assumed in point 3, reduce its probability of occurrence, or initiate an alternate operational mode which allows continued operation when the failure occurs.
8. Evaluation of the severity level of the failure mode assumed in point 3 (usual classification: 1 = minor, 2 = major, 3 = critical, 4 = catastrophic).
9. Estimation of the probability of occurrence (or of the associated failure rate) of the failure mode assumed in point 3, under consideration of the causes identified in point 4.
10. Formulation of pertinent remarks which complete the information in the previous columns, and also of recommendations for corrective actions which will reduce the consequences of the failure mode assumed in point 3.

Table 5. **Procedure for performing a failure modes, effects, and criticality analysis (FMECA)**

design reviews, qualification and production tests, corrective actions, and a reliability growth program [1,8-10,31,41,50,52,66].

CHAPTER 3

STOCHASTIC PROCESSES USED IN MODELING RELIABILITY PROBLEMS

The investigation of the time behaviour of a repairable system, and in particular of its reliability and availability, can theoretically always be performed using stochastic processes. However, in order to be analytically tractable, it is generally necessary to confine the reliability models to a class which can be handled with regenerative stochastic processes. These include renewal and alternating renewal processes, Markov processes with a finite state space, semi-Markov processes, and regenerative stochastic processes with only one (or a few) regeneration state(s). Particular models can also be investigated using some kinds of non-regenerative stochastic processes, like superimposed renewal processes and processes with supplementary variables. This chapter introduces the stochastic processes used in the modeling of reliability problems. Emphasis is given to aspects which are important in practical applications (chapters 4 through 6).

3.1 Renewal processes

3.1.1 Definition and general properties

Renewal processes are the simplest regenerative stochastic processes. In reliability theory they describe the model of a renewable item which is instantaneously renewed at each failure [3,7-9,12,14,21,26,45,48,53-55,60,62,63,71]. To mathematically define the renewal process, let $\tau_0, \tau_1, \tau_2, \ldots$ be statistically independent, non-negative random variables (failure-free times in our context), and let

$$F_A(x) = \Pr\{\tau_0 \leqslant x\} \tag{35}$$

and

$$F(x) = \Pr\{\tau_i \leqslant x\}, \qquad i = 1,2,3,\ldots \tag{36}$$

be the distribution functions of τ_0 and of τ_i, respectively. The random variables

$$S_n = \sum_{i=0}^{n-1} \tau_i, \qquad n = 1,2,3,\ldots$$

or equivalently the sequence $\{\tau_i\}$, $i \geqslant 0$, constitute a renewal process. Fig.10 shows a possible time schedule. The points $S_1, S_2, S_3, \ldots$ are *renewal points*. The time origin $t=0$ is a renewal point only if $F_A(x) = F(x)$. The renewal process is a *point process*. Associated with it, is a count function $\zeta(t)$, starting at 0 for $t=0$ and jumping a unit step at each renewal point (Fig.10).

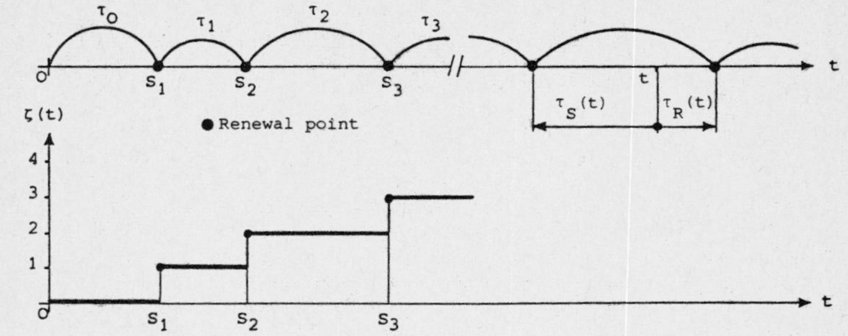

Fig. 10. **Time schedule of a renewal process and of the associated count function $\zeta(t)$**

Renewal processes are *ordinary* if $F_A(x) = F(x)$, *stationary* if

$$F_A(x) = \frac{1}{MTTF} \int_0^x (1-F(y))dy, \tag{37}$$

where

$$MTTF = E\{\tau_i\} = \int_0^\infty (1-F(y))dy, \qquad i \geq 1, \tag{38}$$

and *modified*, otherwise. As shown below (eq.(60)), for a stationary renewal process, the distribution function of the forward recurrence time $(\tau_R(t))$ in Fig.10 is independent of t and equal to $F_A(x)$ given by equation (37). From this, the count function $\zeta(t)$ associated with a stationary renewal process is a time-homogeneous stochastic process.

To simplify the investigation let us assume that

1. $F_A(0) = F(0) = 0.$ (39)
2. The densities $f_A(x) = \frac{dF_A(x)}{dx}$ and $f(x) = \frac{dF(x)}{dx}$ exist for $x > 0$ and go to 0 as $x \to \infty$. (40)
3. $E\{\tau_i\}$ and $Var\{\tau_i\}$ exist for $i \geq 0$. (41)

3.1.2 Renewal function and renewal density

Consider first the expected number of renewal points in the time interval $(0,t)$. From Fig.10 one has

$$\Pr\{\zeta(t) \leq n-1\} = \Pr\{S_n > t\} = 1-\Pr\{S_n \leq t\}$$
$$= 1-\Pr\{\tau_0+\tau_1+\tau_2+\ldots+\tau_{n-1} \leq t\}$$
$$= 1-F_n(t), \qquad n = 1,2,3,\ldots \; . \tag{42}$$

The functions $F_n(t)$ are evaluated recurrently using

$$F_1(t) = F_A(t),$$
$$F_{n+1}(t) = \int_0^t F_n(t-x)f(x)dx, \qquad n = 1,2,3,\ldots \; . \tag{43}$$

From equation (42) it follows that

$$\Pr\{\zeta(t) = n\} = \Pr\{\zeta(t) \leq n\} - \Pr\{\zeta(t) \leq n-1\}$$
$$= F_n(t)-F_{n+1}(t), \quad n = 1,2,3,\ldots \tag{44}$$

and finally

$$E\{\zeta(t)\} = \sum_{n=1}^{\infty} n[F_n(t)-F_{n+1}(t)]$$
$$= \sum_{n=1}^{\infty} F_n(t) = H(t). \tag{45}$$

The function $H(t)$ is called the *renewal function.* Its derivative

$$h(t) = \frac{dH(t)}{dt} = \sum_{n=1}^{\infty} f_n(t) \tag{46}$$

is the *renewal density.* To better understand the meaning of $h(t)$, let us consider the increment of $H(t)$. From equation (45) it follows that

$$\delta H(t) = H(t+\delta t)-H(t) = \sum_{n=1}^{\infty} (F_n(t+\delta t)-F_n(t))$$
$$= \sum_{n=1}^{\infty} \Pr\{t < S_n \leq t+\delta t\}. \tag{47}$$

On the other hand, because of equations (39) and (40), one has for $\delta t \to 0$

$$\Pr\{\text{more than one renewal point in } (t,t+\delta t)\} \to o(\delta t). \tag{48}$$

For $\delta t \to 0$, equation (47) leads then to

$$\delta H(t) = \Pr\{S_1 \text{ or } S_2 \text{ or } S_3 \text{ or} \ldots \text{lies in } (t,t+\delta t)\} \to h(t)\,\delta t. \tag{49}$$

The interpretation of the renewal density given by equation (49) is important in practical applications. It must be pointed out that the renewal density $h(t)$ given by equation (46) differs fundamentally from the failure rate $\lambda(t)$ defined by equation (3) - even in the case of the Poisson process, for which $\lambda(t) = h(t) = \lambda$ holds (see eq. (7) and (65)).

The Laplace transform (eq.(28)) of the renewal density is given by

$$\tilde{h}(s) = \sum_{n=1}^{\infty} \tilde{f}_n(s) = \tilde{f}_A(s)+\tilde{f}_A(s)\tilde{f}(s)+\tilde{f}_A(s)\tilde{f}(s)\tilde{f}(s)+\ldots = \frac{\tilde{f}_A(s)}{1-\tilde{f}(s)}. \quad (50)$$

From equation (50) it can be easily verified that $h(t)$ satisfies the following integral equation

$$h(t) = f_A(t)+\int_0^t h(y)f(t-y)dy. \quad (51)$$

Integral equations of this kind will occur frequently in the next sections and chapters (see e.g. eq.(151) for a derivation).

A practical application of equations (42) and (43) is the determination of the number of spare parts necessary to cover, with probability $\geqslant\gamma$, replacement requirements during the operational time T_0. Assuming the spare parts are non-repairable and their failure-free times (life times) are distributed according to $F(x)$ with mean MTTF, the problem is to find the smallest value of n for which

$$\Pr\{\tau_1+\tau_2+\ldots+\tau_n > T_0\} \geqslant \gamma \quad (52)$$

holds. An analytical solution can be found if $F(x)$ is a gamma distribution or a special case of it (exponential, Erlang or χ^2). For the Weibull distribution

$$F(x) = 1-e^{-(\lambda x)^{\beta}} \quad (53)$$

with $\beta > 1$, i.e. for parts subjected to wear-out and/or fatigue, numerical solutions for $\gamma = 0.99$ and 0.95, and for $\beta = 1, 1.2, 1.5, 2, 3$ and 4 are given in Fig.11 [237]. Also shown in Fig.11 (dashed lines) are the solutions obtained using the central limit theorem, i.e. from

$$n = [d/2+\sqrt{(d/2)^2+T_0/\mathrm{MTTF}}]^2 \quad (54)$$

with $d = B\cdot 2.33$ for $\gamma = 0.99$ and $d = B\cdot 1.64$ for $\gamma = 0.95$, and

$$B = \frac{\sqrt{\mathrm{Var}\{\tau_i\}}}{\mathrm{MTTF}} = \left[\frac{\Gamma(1+2/\beta)}{(\Gamma(1+1/\beta))^2} - 1\right]^{1/2}. \quad (55)$$

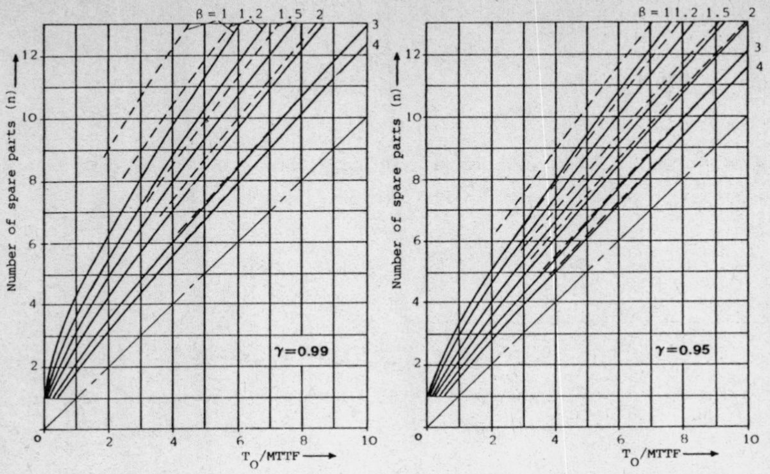

Fig. 11. Number (n) of spare parts which are necessary to assure, with probability $\geq \gamma$, operation over $(0,T_0)$ $(\gamma = \Pr\{\tau_1+\tau_2+\ldots+\tau_n > T_0\}$ with $\Pr\{\tau_i \leq x\} = 1-e^{-(\lambda x)^\beta})$

3.1.3 Forward and backward recurrence-times

Let us now consider the distribution of the forward and the backward recurrence-times. As forward recurrence-time, also known as residual waiting time, one designates the random time interval $\tau_R(t)$ from an arbitrary time point t forward to the next renewal point (Fig.10). To find the distribution function of $\tau_R(t)$, consider that the event $\tau_R(t) > x$ occurs with exactly one of the following mutually exclusive events:

- $A_0 = S_1 > t+x$
- $A_n = (S_n \leq t) \cap (\tau_n > t+x-S_n)$, $\qquad n = 1,2,3,\ldots$.

The event A_n means that exactly n renewal points have occurred before t and the (n+1)th renewal point occurs after t+x. For the event $\tau_R(t) > x$, it then follows that

$$\Pr\{\tau_R(t) > x\} = 1-F_A(t+x)+\int_0^t h(y)(1-F(t+x-y))dy,$$

and finally

$$\Pr\{\tau_R(t) \leq x\} = F_A(t+x)-\int_0^t h(y)(1-F(t+x-y))dy. \qquad (56)$$

As backward recurrence-time, also known as spent waiting time, one designates the random time interval $\tau_S(t)$ from an arbitrary time point t backward to the last renewal point or to the time origin if $S_1 > t$. Similarly, as with $\tau_R(t)$, one obtains

$$\Pr\{\tau_S(t) \leq x\} = \begin{cases} \int_{t-x}^{t} h(y)(1-F(t-y))dy & \text{for } x < t \\ 1 & \text{for } x \geq t. \end{cases} \quad (57)$$

The distribution function of $\tau_S(t)$ has a jump of height $1-F_A(t) = \Pr\{S_1 > t\}$ at the point $x = t$.

3.1.4 Asymptotic and stationary behaviour

The asymptotic behaviour of renewal processes, and in particular of the distribution functions of $\tau_R(t)$ and $\tau_S(t)$, is investigated using two fundamental theorems of renewal theory:

1. The *renewal density theorem* [18,53,54]: If $f_A(x)$ and $f(x)$ satisfy the conditions stated with equations (39) to (41), then

   $$\lim_{t\to\infty} h(t) = \frac{1}{MTTF} \quad (58)$$

   holds with MTTF as in equation (38).
2. The *key renewal theorem* [19,53]: If $U(z)$ is directly Riemann integrable [19]; and in particular, if $U(z)$ is ≥ 0, non-increasing, and Riemann integrable over $(0,\infty)$, then

   $$\lim_{t\to\infty} \int_0^t U(t-y)h(y)dy = \frac{1}{MTTF}\int_0^\infty U(z)dz \quad (59)$$

   holds with MTTF as in equation (38).

With the help of equations (58) and (59) it follows that

$$\lim_{t\to\infty} \Pr\{\tau_R(t) \leq x\} = \frac{1}{MTTF}\int_0^x (1-F(y))dy \quad (60)$$

and

$$\lim_{t\to\infty} \Pr\{\tau_S(t) \leq x\} = \frac{1}{MTTF}\int_0^x (1-F(y))dy. \quad (61)$$

A comparison of equations (60) and (37) shows the strong relation existing between asymptotic behaviour and stationary state. This leads to the following interpretation, useful for checking *stationarity* in practical applications:

> A stationary renewal process can be considered as an ordinary or a modified renewal process which has been started at $t = -\infty$ and which will be observed for time $t \geq 0$,

$t = 0$ being an arbitrary time point. Such an interpretation holds in general for all processes discussed in this chapter. A summary of the main properties of stationary renewal processes is given in Table 6.

Quantity	Expression	Remarks, assumptions
1. Distribution function of τ_0	$F_A(x) = \frac{1}{MTTF} \int_0^x (1-F(y))dy$	$f_A(x) = dF_A(x)/dx, \quad x>0$ $MTTF = E\{\tau_i\}, \quad i \geq 1$
2. Distribution function of τ_i $(i \geq 1)$	$F(x)$	$F(0) = 0;$ $f(x) = dF(x)/dx, \quad x>0$
3. Renewal function	$H(t) = \frac{t}{MTTF}$	$H(t) = E\{$number of renewal points in $(0,t)\}$
4. Renewal density	$h(t) = \frac{1}{MTTF}$	$h(t) = dH(t)/dt;$ $Pr\{S_1$ or S_2 or S_3 or... lies in $(t,t+\delta t)\} \to h(t)\delta t$
5. Distribution function of the forward recurrence time	$Pr\{\tau_R(t) \leq x\} = F_A(x)$	independent of t; $F_A(x)$ as in point 1

Table 6. **Main properties of stationary renewal processes**

3.1.5 Poisson process

An important renewal process is the (homogeneous) Poisson process. This process is characterized by

$$F_A(x) = F(x) = 1-e^{-\lambda x}. \tag{62}$$

From the above results it follows that

$$F_n(t) = 1- \sum_{i=0}^{n-1} \frac{(\lambda t)^i}{i!} e^{-\lambda t}, \qquad n = 1,2,3,... \tag{63}$$

$$H(t) = \lambda t \tag{64}$$

$$h(t) = \lambda \tag{65}$$

$$Pr\{\tau_R(t) \leq x\} = 1-e^{-\lambda x} \tag{66}$$

$$Pr\{\tau_S(t) \leq x\} = \begin{cases} 1-e^{-\lambda x} & \text{for } x<t \\ 1 & \text{for } x \geq t. \end{cases} \tag{67}$$

The Poisson process is stationary. It is the only continuous-time renewal process for which the forward recurrence time $\tau_R(t)$ is distributed according to the same exponential distribution as all random time intervals $\tau_i (i \geq 0)$, irrespective of both t and the initial conditions at $t = 0$. This confirms the *memoreyless* property of the exponential distribution (eq. (62)), discussed in section 2.2.

3.2 Alternating renewal processes

Generalizing the renewal process of Fig.10 by introducing a random replacement time at each failure, distributed say according to $G(x)$, one obtains the alternating renewal process [3,7-9,14,21,26,48,60,62,71]. The alternating renewal process is a two state process which alternates from one state to the other after a sojourn time distributed according to $F(x)$ and $G(x)$, respectively. Considering the practical applications of chapters 4 trough 6, the two states will be called up and down, and abbreviated as u and d. To mathematically define the alternating renewal process, let $\{\tau_i\}$ and $\{\tau'_i\}$ be two statistically independent renewal processes characterized by the distribution functions

$$F_A(x) \text{ for } \tau_0 \text{ and } F(x) \text{ for } \tau_i, \quad i \geq 1$$
$$G_A(x) \text{ for } \tau'_0 \text{ and } G(x) \text{ for } \tau'_i, \quad i \geq 1,$$

with densities $f_A(x), f(x), g_A(x), g(x)$ and means

$$MTTF = E\{\tau_i\} \quad \text{and} \quad MTTR = E\{\tau'_i\}, \qquad i \geq 1, \tag{68}$$

where MTTF stand for *mean time-to-failure* and MTTR for *mean time-to-repair*. It is assumed that the conditions stated with equations (39) to (41) hold. The sequences

$$\tau_0, \tau'_1, \tau_1, \tau'_2, \tau_2, \tau'_3, \ldots$$

and

$$\tau'_0, \tau_1, \tau'_1, \tau_2, \tau'_2, \tau_3, \ldots$$

(obtained by taking alternately one random time interval of type τ and one of Type τ') constitute two modified alternating renewal processes starting at $t=0$ with τ_0 and τ'_0 respectively, see Fig.12. In each of these alternating renewal processes it is possible to define two *embedded renewal processes* with renewal points S_{udui} and S_{duui} for the alternating renewal process starting up at $t=0$, and S_{uddi} and S_{dudi} for that one starting down at $t=0$ (Fig.12). The embedded renewal processes are modified and differ from each other only in the first random time intervals. These are for the four cases

$$\tau_0, \tau_0+\tau'_1, \tau'_0+\tau_1, \tau'_0,$$

the corresponding densities being

$$f_A(x), f_A(x)*g(x), g_A(x)*f(x), g_A(x),$$

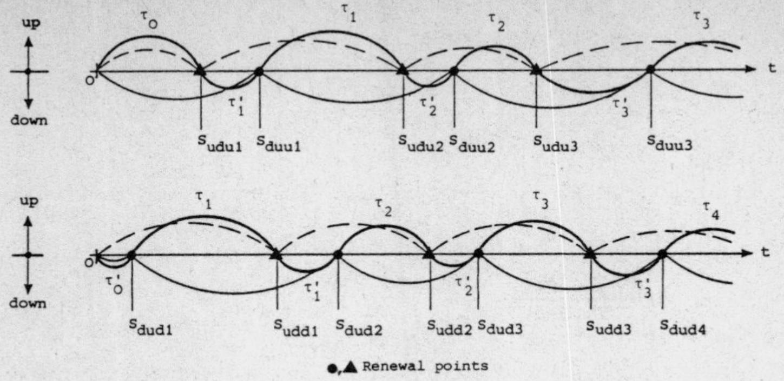

Fig. 12. **Time schedule of two alternating renewal processes starting up, or down, at t = 0 (also shown are the embedded renewal processes with renewal points ● and ▲)**

where the symbol $*$ denotes convolution

$$f_A(x)*g(x) = \int_0^x f_A(y)g(x-y)dy. \tag{69}$$

All other random time intervals have a density $f(x)*g(x)$.

The results of the preceding section can be used to investigate the alternating renewal processes of Fig.12. In particular, one can define the renewal densities

$$h_{udu}(t), h_{duu}(t), h_{udd}(t), h_{dud}(t)$$

whose Laplace transforms are (eq.(50))

$$\tilde{h}_{udu}(s) = \frac{\tilde{f}_A(s)}{1-\tilde{f}(s)\tilde{g}(s)}, \qquad \tilde{h}_{duu}(s) = \frac{\tilde{f}_A(s)\tilde{g}(s)}{1-\tilde{f}(s)\tilde{g}(s)},$$
$$\tilde{h}_{udd}(s) = \frac{\tilde{g}_A(s)\tilde{f}(s)}{1-\tilde{f}(s)\tilde{g}(s)}, \qquad \tilde{h}_{dud}(s) = \frac{\tilde{g}_A(s)}{1-\tilde{f}(s)\tilde{g}(s)}. \tag{70}$$

Furthermore, the renewal density theorem (eq.(58)) and the key renewal theorem (eq.(59)) hold for all four embedded renewal processes, with

MTTF+MTTR instead of MTTF,

thus allowing the investigation of asymptotic behaviour (section 4.5).

The two processes of Fig.12 can also be combined by defining

$$p = \Pr\{\text{up at } t=0\} \quad \text{and} \quad 1-p = \Pr\{\text{down at } t=0\}. \tag{71}$$

In a way similar to the renewal processes, alternating renewal processes are *ordinary* if $F_A(x) = F(x)$ and $G_A(x) = G(x)$, *stationary* if

$$p = \frac{MTTF}{MTTF+MTTR}, \quad F_A(x) = \frac{1}{MTTF}\int_0^x (1-F(y))dy, \quad G_A(x) = \frac{1}{MTTR}\int_0^x (1-G(y))dy \tag{72}$$

and *modified*, otherwise. For the stationary state one obtains in particular (section 4.6)

$$\Pr\{\text{up at the time } t\} = \frac{MTTF}{MTTF+MTTR} = PA \tag{73}$$

and

$$PA\, h_{udu}(t) + (1-PA)\, h_{udd}(t) = PA\, h_{duu}(t) + (1-PA)\, h_{dud}(t) = \frac{1}{MTTF+MTTR} \tag{74}$$

for all $t \geq 0$. As with renewal processes, the following interpretation is useful for checking *stationarity* in practical applications:

> A stationary alternating renewal process can be considered as an ordinary or modified alternating renewal process which has been started at $t = -\infty$ and which will be observed for times $t \geq 0$,

$t = 0$ being an arbitrary time point.

3.3 Markov processes with a finite state space

3.3.1 Definition and general properties

A stochastic process $\xi(t)$ with state space $Z_0, Z_1, \ldots, Z_m$ is a Markov process if, given the state occupied at a time t, say Z_i, the future development of the process depends on Z_i and t, but not on the process development up to the time t. For such a process the relation

$$\Pr\{\xi(t+a) = Z_j \mid (\xi(t) = Z_i \cap \xi(t_n) = Z_{i_n} \cap \xi(t_{n-1}) = Z_{i_{n-1}} \cap \ldots \cap \xi(t_1) = Z_{i_1})\} = \Pr\{\xi(t+a) = Z_j \mid \xi(t) = Z_i\} \tag{75}$$

holds for any $n = 1,2,3,\ldots$, for all $t+a > t > t_n > \ldots > t_1$, and for any $i, j, i_1, i_2, \ldots, i_n$ $(i, j, i_1, i_2, \ldots, i_n = 0,1,\ldots,m)$. The conditional probabilities given by equation (75) are the *transition probabilities* $P_{ij}(t, t+a)$ of the Markov process

$$P_{ij}(t, t+a) = \Pr\{\xi(t+a) = Z_j \mid \xi(t) = Z_i\}. \tag{76}$$

In the following it is assumed that

$$0 \leq p_{ij}(t,t+a) \leq 1$$

and

$$\sum_{j=0}^{m} p_{ij}(t,t+a) = 1$$

hold for all t,t+a (stochastic matrix). Together with the *initial conditions*

$$P_i(0) = \Pr\{\xi(0) = Z_i\}, \qquad i = 0,1,\ldots,m \tag{77}$$

the transition probabilities $p_{ij}(t,t+a)$ completely determine the stochastic behaviour of the Markov process. For instance, the *state probabilities*

$$P_i(t) = \Pr\{\xi(t) = Z_i\}, \qquad i = 0,1,\ldots,m, \tag{78}$$

are given by

$$P_i(t) = \sum_{k=0}^{m} P_k(0) p_{ki}(0,t), \tag{79}$$

with

$$\sum_{i=0}^{m} P_i(t) = 1. \tag{80}$$

3.3.2 Transition rates

Markov processes have been widely considered in the literature [3,4,9,12,17,19,21,29, 45,48,49,60,65]. In view of the practical applications of chapters 4 through 6, the investigations given here will be limited to the *time-homogeneous* Markov processes, for which the transition probabilities are independent of t

$$p_{ij}(t,t+a) = p_{ij}(a), \tag{81}$$

and make use of the following *semi-Markov transition probabilities* (see eq.(120))

$$Q_{ij}(x) = \Pr\{(\text{sojourn time in } Z_i \leq x \cap \text{next transition is in } Z_j) \mid Z_i \text{ is entered at } x = 0\}. \tag{82}$$

A simple transformation leads to

$$Q_{ij}(x) = p_{ij} F_{ij}(x), \tag{83}$$

where

$$p_{ij} = \Pr\{\text{next transition is in } Z_j \mid Z_i \text{ is entered at } x=0\} \tag{84}$$

are the transition probabilities of a *Markov chain embedded in the Markov process*, with

$$p_{ii} = 0 \quad \text{and} \quad \sum_{j=0}^{m} p_{ij} = 1, \tag{85}$$

and

$$F_{ij}(x) = \Pr\{\text{sojourn time in } Z_i \leqslant x \mid (\text{next transition is in } Z_j \cap Z_i \text{ is entered at } x=0)\} \tag{86}$$

are the conditional distribution functions of the sojourn times in state Z_i. (It is also possible to define embedded Markov chains for which $p_{ii} \geqslant 0$ holds. This will be in general necessary in the case of regenerative stochastic processes with only a few regeneration states, see section 3.5.) In practical applications, the following interpretation is useful for determining the semi-Markov transition probabilities $Q_{ij}(x)$:

> Assume that a transition in the state Z_i occurs just now $(x=0)$; at this time point the sojourn times $\tau_{i0}, \tau_{i1}, \ldots, \tau_{im}$ (without τ_{ii}) are started; the next transition will be x units of time later in the state Z_j if one has $\tau_{ij} = x$ and $\tau_{ik} > \tau_{ij}$ for all $k \neq j$. In this model, the quantities $Q_{ij}(x)$, p_{ij} and $F_{ij}(x)$ have the following meaning:

$$Q_{ij}(x) = \Pr\{\tau_{ij} \leqslant x \cap \tau_{ik} > \tau_{ij},\ \ k \neq j\} \tag{87}$$

$$p_{ij} = \Pr\{\tau_{ik} > \tau_{ij},\ \ k \neq j\} \tag{88}$$

$$F_{ij}(x) = \Pr\{\tau_{ij} \leqslant x \mid \tau_{ik} > \tau_{ij},\ \ k \neq j\}. \tag{89}$$

Because of the *memoryless* property of the time-homogeneous Markov process, stated with equations (75) and (81), all τ_{ij} are exponentially distributed (section 3.1.5). Let

$$\Pr\{\tau_{ij} \leqslant x\} = 1 - e^{-\rho_{ij} x}.$$

From equation (87) it follows that

$$Q_{ij}(x) = \int_0^x \rho_{ij} e^{-\rho_{ij} y} e^{-\sum_{k \neq j} \rho_{ik} y} dy = \frac{\rho_{ij}}{\rho_i} (1 - e^{-\rho_i x}), \tag{90}$$

with

$$\rho_i = \sum_{j=0}^{m} \rho_{ij}, \quad \rho_{ij} \geqslant 0, \quad \rho_{ii} = 0, \tag{91}$$

and consequently

$$p_{ij} = \frac{\rho_{ij}}{\rho_i} \tag{92}$$

and

$$F_{ij}(x) = 1-e^{-\rho_i x}. \tag{93}$$

The parameters ρ_{ij} are the *transition rates* of the time-homogeneous Markov process. They can be obtained either from the transition probabilities $p_{ij}(a)$

$$\rho_{ij} = \lim_{a\to 0} \frac{p_{ij}(a)}{a} \tag{94}$$

$$\rho_i = \lim_{a\to 0} \frac{1-p_{ii}(a)}{a}, \tag{95}$$

or from equations (82) and (90) as limit for $x\to 0$, by considering that $\lim_{\delta t\to 0}$ Pr{more than one transition in $(t,t+\delta t)\}\to o(\delta t)$,

$$\rho_{ij} = \lim_{\delta t\to 0} \frac{1}{\delta t} \Pr\{\text{transition from } Z_i \text{ to } Z_j \text{ in } (t,t+\delta t) \mid \xi(t) = Z_i\} \tag{96}$$

$$\rho_i = \lim_{\delta t\to 0} \frac{1}{\delta t} \Pr\{\text{transition from } Z_i \text{ to any other state in } (t,t+\delta t) \mid \xi(t) = Z_i\}. \tag{97}$$

The transition rates have a great intuitive appeal and can be used to graphically visualize the behaviour of the time-homogeneous Markov process in an arbitrary time interval $(t,t+\delta t)$, as with the transition probabilities p_{ij} of a homogeneous Markov chain.

As an example let us consider a 1-out-of-2 warm redundancy with failure-free times distributed according to $F(x) = 1-e^{-\lambda x}$ in the operating state and $F_r(x) = 1-e^{-\lambda_r x}$ in the reserve state, repair times distributed according to $G(x) = 1-e^{-\mu x}$, and with one repair crew. Repair is started as soon as a failure occurs and the repair crew is not affected by pevious failures. Repair of a redundant element has no influence on the element in the operating state. Because of the constant failure and repair rates (λ,λ_r,μ), the stochastic process involved is a time-homogeneous Markov process with the states Z_0, Z_1 and Z_2, where the indices correspond to the number of elements in the down state. At state Z_2 the system is down. Fig.13 shows the transition probabilities diagram for an arbitrary time interval $(t,t+\delta t)$, and gives the corresponding transition rates. Considering the process at two adjacent time points t and $t+\delta t$, as in section 2.4, Fig.13 leads to the following difference equations for state probabilities $P_0(t)$, $P_1(t)$ and $P_2(t)$:

$$\begin{aligned}
P_0(t+\delta t) &= P_0(t)(1-(\lambda+\lambda_r)\delta t)+P_1(t)\mu\delta t\\
P_1(t+\delta t) &= P_1(t)(1-(\lambda+\mu)\delta t)+P_0(t)(\lambda+\lambda_r)\delta t+P_2(t)\mu\delta t\\
P_2(t+\delta t) &= P_2(t)(1-\mu\delta t)+P_1(t)\lambda\delta t.
\end{aligned} \tag{98}$$

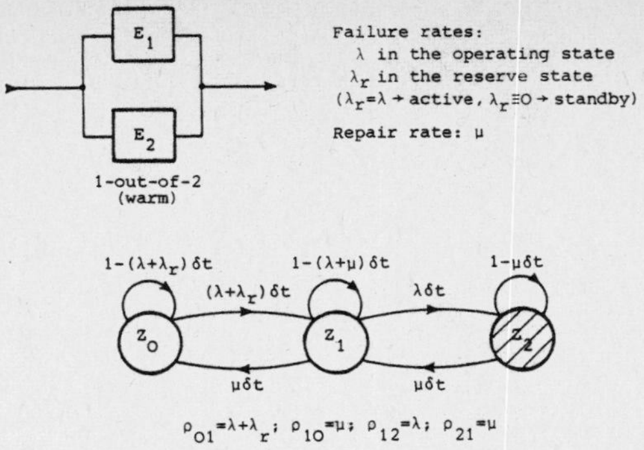

Fig. 13. Diagram of the transition probabilities in $(t,t+\delta t)$, and transition rates (ρ_{ij}), for a repairable 1-out-of-2 warm redundancy (constant failure and repair rates, only one repair crew, no further failure at system down, t arbitrary, $\delta t \to 0$)

For $\delta t \to 0$ it then follows that

$$\begin{aligned}
\dot{P}_0(t) &= -(\lambda+\lambda_r)P_0(t) + P_1(t)\mu \\
\dot{P}_1(t) &= -(\lambda+\mu)P_1(t) + P_0(t)(\lambda+\lambda_r) + P_2(t)\mu \\
\dot{P}_2(t) &= -\mu P_2(t) + P_1(t)\lambda
\end{aligned} \tag{99}$$

As a further example, Fig.14a shows the transition probabilities diagram for an arbitrary time interval $(t,t+\delta t)$ and gives the transition rates for the case of a 1-out-of-2 active redundancy with different elements and only one repair crew. This model can also be used to describe the 1-out-of-2 redundancy of Fig.13 in the case where failures in the operating and reserve states must be considered separately. The situation of a 1-out-of-2 active redundancy with identical elements and a switch in series is illustrated in Fig.14b. It is here assumed that no further failure can occur at system down.

3.3.3 State probabilities

The procedure used to determine equations (98) and (99) can be generalized to an arbitrary structure and leads to the following system of differential equations for the state probabilities $P_j(t)$:

$$\dot{P}_j(t) = -\rho_j P_j(t) + \sum_{k=0}^{m} P_k(t)\rho_{kj}, \qquad j = 0,1,\dots,m \tag{100}$$

with ρ_j and ρ_{kj} as defined by equations (91), (96) and (97). Assuming as initial conditions

$$P_i(0) = 1, \quad P_j(0) = 0 \quad \text{for} \quad j \neq i, \tag{101}$$

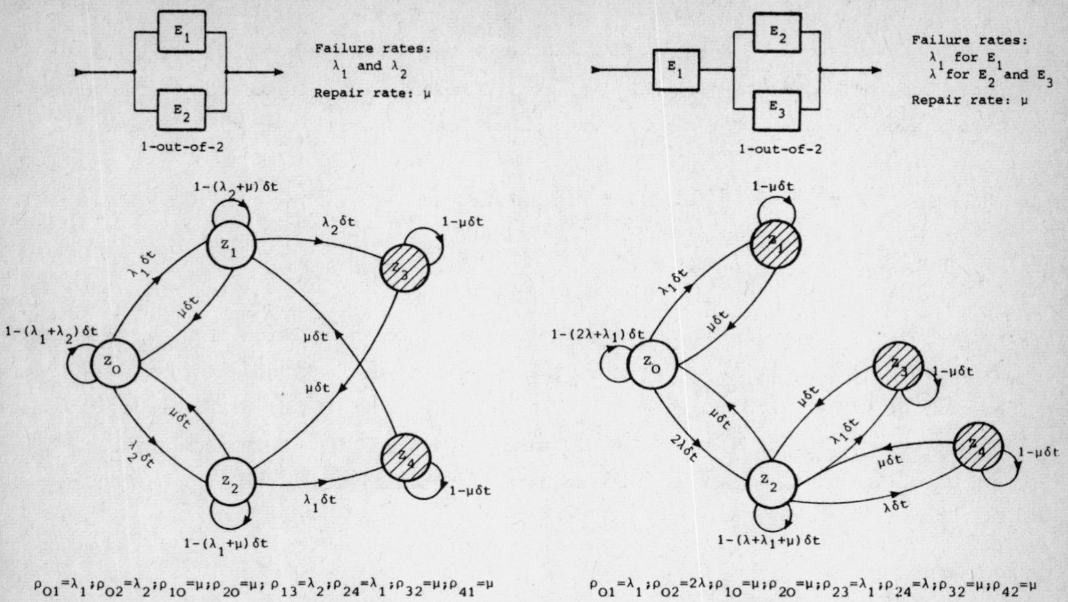

(a) Different elements without switch

(b) Identical elements with switch in series

Fig. 14. Diagrams of the transition probabilities in $(t,t+\delta t)$, and transition rates (ρ_{ij}), for repairable 1-out-of-2 active redundancies (constant failure and repair rates, only one repair crew, no further failure at system down, t arbitrary, $\delta t \to 0$)

the solutions of equations (100) and (101) leads to the *conditional state probabilities* $P_{ij}(t)$, defined as

$$P_{ij}(t) = \Pr\{\xi(t) = Z_j \mid Z_i \text{ is entered at } t = 0\} \qquad i,j = 0,1,\dots,m \tag{102}$$

and obtained by setting $P_{ij}(t) = P_j(t)$, with $P_j(t)$ from equations (100) and (101).

In reliability applications, it is useful to split the state space into two complementary sets

$$\begin{aligned} M &= \text{set of the up states} \\ \bar{M} &= \text{set of the down states.} \end{aligned} \tag{103}$$

The probability of being in M at the time t, given that the state Z_i is entered at $t = 0$, is the *point-availability* $PA_{Si}(t)$

$$PA_{Si}(t) = \Pr\{\text{system up at } t \mid Z_i \text{ is entered at } t = 0\}, \quad i = 0,1,\dots,m \tag{104}$$

which can be computed using equation (102)

$$PA_{Si}(t) = \sum_{Z_j \in M} P_{ij}(t) \qquad i = 0,1,\dots,m. \tag{105}$$

The functions $P_{ij}(t)$ as defined above cannot be used to compute the reliability function, because they do not consider that each transition from M to $\bar{M}$ is a system failure. The *reliability function* $R_{Si}(t)$ is the probability that the first transition from M to $\bar{M}$ occurs after the time t, given that $Z_i \in M$ is entered at $t=0$

$$R_{Si}(t) = \Pr\{\text{system up in } (0,t) \mid Z_i \text{ is entered at } t=0\}, \quad Z_i \in M. \tag{106}$$

It can be computed in two different ways:

1. By making all states in $\bar{M}$ *absorbing*, i.e. by setting in equation (100) $\rho_{kj}=0$ for all $Z_k \in \bar{M}$ $(j=0,1,\ldots,m)$. As long as the process moves in M, the stochastic behaviour of the modified process is identical to that of the original process. If the process enters in $\bar{M}$, then the modified process remains there indefinitely. To avoid confusion, the state probabilities of the modified process will be designated with $P'_j(t)$ instead of $P_j(t)$. Equations (100) and (101) then become

$$\dot{P}'_j(t) = -\rho_j P'_j(t) + \sum_{Z_k \in M} P'_k(t)\rho_{kj}, \tag{107}$$

and

$$P'_i(0) = 1, \quad P'_j(0) = 0 \quad \text{for } j \neq i, \qquad Z_i \in M. \tag{108}$$

The solution of equations (107) and (108) leads to the *conditional state probabilities for the modified process* $P'_{ij}(t)$, defined as

$$P'_{ij}(t) = \Pr\{(\xi(t) = Z_j \cap \text{in M during } (0,t)) \mid Z_i \text{ is entered at } t=0\}, \quad Z_i, Z_j \in M \tag{109}$$

and obtained by putting $P'_{ij}(t) = P'_j(t)$ with $P'_j(t)$ from equations (107) and (108). From this it follows that

$$R_{Si}(t) = \sum_{Z_j \in M} P'_{ij}(t), \qquad Z_i \in M. \tag{110}$$

2. By computing directly the probability that the time to the first transition from M to $\bar{M}$ is greater than t, given $Z_i \in M$ is entered at $t=0$. This probability ($R_{Si}(t)$) is the solution of the following system of integral equations:

$$R_{Si}(t) = e^{-\rho_i t} + \sum_{Z_j \in M} \int_0^t \rho_{ij} e^{-\rho_i x} R_{Sj}(t-x)\,dx, \qquad Z_i \in M. \tag{111}$$

Because of the *memoryless* property of the Markov process, the condition *Z_i is entered at $t=0$* can be substituted by *the process is in Z_i at $t=0$*. The first term of equation

(111) gives the probability that the process will not leave the state Z_i before the time t. The second term takes care of the fact that if a transition from the state Z_i to the state $Z_j \in M$ occurs at the time $x \leq t$, then at this transition point the process finds itself, with respect to $R_{Sj}(t-x)$, in the same situation as at the time $t=0$ with respect to $R_{Si}(t)$. The corresponding probability must be summed over all $x \leq t$ and over all states $Z_j \in M$. A similar derivation will be given in section 4.2 for the point-availability $PA_u(t)$ of a one-item repairable structure (eq. (151)).

3.3.4 Asymptotic and stationary behaviour

Although theoretically possible, the evaluation for large systems of the state probabilities (eq. (100)) or of the reliability functions (eq. (111)) can require a great amount of resources. This effort can be avoided if one is merely interested in asymptotic and stationary results or in mean values like mean up-time, mean down-time, etc.

To investigate the asymptotic and stationary behaviour, let us assume that the embedded Markov chain is *irreducible* (every state can be reached from every other state with probability >0). For such a process the limits

$$\lim_{t \to \infty} P_j(t) = \lim_{t \to \infty} P_{ij}(t) = p_j, \qquad j = 0, 1, \ldots, m \tag{112}$$

exist, with

$$p_j > 0 \quad \text{and} \quad \sum_{j=0}^{m} p_j = 1, \tag{113}$$

irrespective of i (ergodic Markov process). p_j are obtained from equation (100) by putting $\dot{P}_j(t) = 0$ and $P_j(t) = p_j$, i.e. by solving

$$\rho_j p_j = \sum_{k=0}^{m} p_k \rho_{kj} \qquad \text{with} \quad \sum_{j=0}^{m} p_j = 1. \tag{114}$$

Taking $P_i(0) = p_i$ for $i = 0, 1, \ldots, m$ as initial conditions in equation (101), the (time-homogeneous) Markov process is *stationary*. In particular it follows then that

$$P_i(t) = p_i \tag{115}$$

for all $t \geq 0$. Furthermore, the expected fraction of the time spent in a given set K of states is given by

$$\frac{1}{t_2 - t_1} \, E\{\text{time spent in K during } (t_1, t_2)\} = \sum_{Z_i \in K} p_i, \tag{116}$$

and expressions of the form $\sum kp_k$ can be used to compute the expected number of elements down, of repair crews in service, of elements in the reserve state, etc.

Also important for reliability analyses are the mean values of some sojourn times up to a given transition. With these belong the mean sojourn time in the state Z_i, given by

$$E\{\text{sojourn time in } Z_i\} = \frac{1}{\rho_i}, \tag{117}$$

and the mean value of the sojourn time in a set M of states up to a transition in $\bar{M}$. This last value is the *mean time-to-system-failure* $MTTF_{Si}$

$$MTTF_{Si} = E\{\text{sojourn time in } M \mid Z_i \text{ is entered at } t=0\}, \qquad Z_i \in M, \tag{118}$$

and can be computed from $MTTF_{Si} = \tilde{R}_{Si}(0)$, where $\tilde{R}_{Si}(0)$ is the Laplace transform of $R_{Si}(t)$ for $s=0$. Using equations (111), (28) and (30), $MTTF_{Si}$ can be obtained by solving the following system of algebraic equations:

$$MTTF_{Si} = \frac{1}{\rho_i} + \sum_{Z_j \in M} \frac{\rho_{ij}}{\rho_i} MTTF_{Sj}, \qquad Z_i \in M. \tag{119}$$

3.3.5 Summary of important relations for Markov models

A summary of the main properties of the time-homogeneous Markov processes with a finite state space is given in Table 7.

3.4 Semi-Markov processes with a finite state space

3.4.1 Definition and general properties

A stochastic process with a finite state space $Z_0, Z_1, \ldots, Z_m$ is a semi-Markov process if it possesses the following property: Given that the state Z_i has just been entered ($x=0$), the sojourn time in Z_i with a consequent jump to the next state to be visited, say Z_j, is a positive random variable τ_{ij} whose distribution function depends on Z_i and Z_j but not on the previous development of the process up to the transition in state Z_i. In the case of a time-homogeneous Markov process, this distribution function has the form $\Pr\{\tau_{ij} \leq x\} = 1-e^{-\rho_{ij}x}$. For a semi-Markov process it is arbitrary, e.g. equal to $F(x)$ or $G(x)$ in the case of the alternating renewal process of section 3.2 (2-state semi-Markov process). From the above statements it follows that the semi-Markov process is Markovian *only* at the transition points, and that successive

Quantity	Expression	Remarks, assumptions
1. Transition rates	$\rho_{ij} = \lim_{\delta t \to 0} \frac{1}{\delta t} \Pr\{\text{transition } Z_i \to Z_j \text{ in } (t, t+\delta t) \mid \xi(t) = Z_i\}$ $\rho_i = \sum_{j=0}^{m} \rho_{ij}, \quad \rho_{ii} = 0$	$Q_{ij} = \Pr\{\tau_{ij} \leq x \cap \tau_{ik} > \tau_{ij}, k \neq j\}$ $\rho_{ij} = \frac{dQ_{ij}(x)}{dx}\Big\vert_{x=0}$
2. State probabilities	$\dot{P}_j(t) = -\rho_j P_j(t) + \sum_{i=0}^{m} P_i(t)\rho_{ij}$	$P_j(t) = \Pr\{\xi(t) = Z_j\}$, $j = 0,1,\dots,m$
3. Point-availability	$PA_{Si}(t) = \sum_{Z_j \in M} P_{ij}(t)$, $i = 0,1,\dots,m$ $P_{ij}(t) = P_j(t)$ from point 2 for $P_i(0) = 1$	$PA_{Si}(t) = \Pr\{\text{in } M \text{ at } t \mid Z_i \text{ is entered at } t = 0\}$ M = set of the up states, $i = 0,1,\dots,m$
4. Reliability function	$R_{Si}(t) = e^{-\rho_i t} + \sum_{Z_j \in M} \int_0^t \rho_{ij} e^{-\rho_i x} R_{Sj}(t-x)\,dx$	$R_{Si}(t) = \Pr\{\text{not to leave } M \text{ in } (0,t] \mid Z_i \text{ is entered at } t = 0\}$ $Z_i \in M$
5. Stationary-state probabilities	$P_j(t) = p_j$, $j = 0,1,\dots,m$	all states are recurrent states; p_j is obtained by solving $p_j \rho_j = \sum_{i=0}^{m} p_i \rho_{ij}$ with $\sum_{j=0}^{m} p_j = 1$
6. Stationary-state point-availability	$PA_S(t) = PA_S = \sum_{Z_j \in M} p_j$	$PA_S = \Pr\{\text{in } M \text{ at } t, \text{in stationary-state}\}$ M = set of the up states; p_j as in point 5
7. Mean time-to-system-failure	$MTTF_{Si} = \frac{1}{\rho_i} + \sum_{Z_j \in M} \frac{\rho_{ij}}{\rho_i} MTTF_{Sj}$, $Z_i \in M$	$MTTF_{Si} = E\{\text{sojourn time in } M \mid Z_i \text{ is entered at } t = 0\}$ M = set of the up states, $Z_i \in M$

Table 7. **Main properties of time-homogeneous Markov processes with a finite state space**

transitions in a given state constitute a renewal process.

More precisely, let $\zeta_0 = Z_{i_0}$, $\zeta_1 = Z_{i_1}$, $\zeta_2 = Z_{i_2}, \dots$ an ordered sequence of state transitions; for a semi-Markov process the relation

$$\begin{aligned}
&\Pr\{(\text{sojourn time in } Z_i \leq x \cap \zeta_{n+1} = Z_j) \mid (\zeta_n = Z_{i_n} = z_i \text{ is entered at } x = 0 \cap \\
&\quad \zeta_{n-1} = Z_{i_{n-1}} \cap \dots \cap \zeta_0 = Z_{i_0} \cap \text{sojourn time in } Z_{i_{n-1}} = x_{i_{n-1}} \cap \dots \\
&\quad \dots \cap \text{sojourn time in } Z_{i_0} = x_{i_0}\} \\
&= \Pr\{(\text{sojourn time in } Z_i \leq x \cap \zeta_{n+1} = Z_j) \mid \zeta_n = Z_i \text{ is entered at } x = 0\} \qquad (120)
\end{aligned}$$

holds for $n = 1,2,3,\dots$, for any $i,j,i_0,i_1,\dots,i_{n-1}$ $(i,j,i_0,i_1,\dots,i_{n-1} = 0,1,\dots,m)$ and for arbitrary sojourn times $x_{i_0}, x_{i_1}, \dots, x_{i_{n-1}}$. The conditional probabilities given by equation (120) are the *semi-Markov transition probabilities* $Q_{ij}(x)$ defined by equation (82)

$$Q_{ij}(x) = \Pr\{(\text{sojourn time in } Z_i \leq x \cap \text{next transition is in } Z_j) \mid Z_i \text{ is entered at } x = 0\}. \qquad (82)$$

As with equations (83) to (86), $Q_{ij}(x)$ can be expressed in the form

$$Q_{ij}(x) = p_{ij} F_{ij}(x), \tag{83}$$

where p_{ij} are the transition probabilities of a *Markov chain embedded in the semi-Markov process* (with $p_{ii} = 0$), and $F_{ij}(x)$ are the conditional distribution functions of the sojourn time in state Z_i.

The interpretation of the functions $Q_{ij}(x)$ given by equation (87)

$$Q_{ij}(x) = \Pr\{\tau_{ij} \leqslant x \cap \tau_{ik} > \tau_{ij}, \quad k \neq j\} \tag{87}$$

also holds here. It is useful to determine the transition probabilities $Q_{ij}(x)$ in practical applications (chapter 5). In the following it is assumed that all τ_{ij} are >0, continuous, and with $E\{\tau_{ij}\} < \infty$ and $\mathrm{Var}\{\tau_{ij}\} < \infty$. Such a semi-Markov process is entirely characterized by the transition probabilities $Q_{ij}(x)$ and the *initial distribution functions*

$$A_{ij}(x) = \Pr\{\xi(0) = Z_i \cap \text{first transition is in } Z_j \cap \text{rest sojourn time in } Z_i \leqslant x\}. \tag{121}$$

Semi-Markov processes have been investigated in [9,12,21,38,47,48,54,55,63]. For reliability analyses one assumes, in general, that at $t=0$ the process either enters a given state Z_i or is stationary.

3.4.2 At $t=0$ the process enters the state Z_i

If at $t=0$ the process enters state Z_i, the conditional state probabilities $P_{ij}(t)$, defined by equation (102) are the solutions of the following system of integral equations:

$$P_{ij}(t) = \delta_{ij}(1-Q_i(t)) + \sum_{k=0}^{m} \int_0^t dQ_{ik}(x) P_{kj}(t-x), \qquad i,j = 0,1,2,\ldots,m \tag{122}$$

with

$$\delta_{ij} = 0 \quad \text{for} \quad j \neq i \quad \text{and} \quad \delta_{ii} = 1, \tag{123}$$

and

$$Q_i(x) = \sum_{j=0}^{m} Q_{ij}(x). \tag{124}$$

The state probabilities (eq.(78)) follow then from equation (122)

$$P_i(t) = \sum_{k=0}^{m} \Pr\{Z_k \text{ is entered at } t=0\} P_{ki}(t), \qquad i=0,1,\dots,m. \tag{125}$$

Splitting up the state space into two complementary sets, M for the up states and $\bar{M}$ for the down states (eq.(103)), the probability of being in M at time t, given that Z_i is entered at t = 0 (point-availability $PA_{Si}(t)$), is given again by equation (105)

$$PA_{Si}(t) = \sum_{Z_j \in M} P_{ij}(t), \qquad i=0,1,\dots,m. \tag{105}$$

with $P_{ij}(t)$ from equation (122). The probability that the first transition from M to $\bar{M}$ occurs after time t, given that $Z_i \in M$ is entered at t = 0 (reliability function $R_{Si}(t)$, eq.(106)), is the solution of the system of integral equations:

$$R_{Si}(t) = 1-Q_i(t) + \sum_{Z_j \in M} \int_0^t dQ_{ij}(x) R_{Sj}(t-x), \qquad Z_i \in M. \tag{126}$$

Equation (126) is a generalization of equation (111). Because of $Q_{ij}(x)$ being arbitrary, the condition *Z_i is entered at t = 0* is necessary. From equation (126) one can evaluate the mean value of the sojourn time in M, given that $Z_i \in M$ is entered at t = 0, by solving

$$MTTF_{Si} = T_i + \sum_{Z_j \in M} p_{ij} MTTF_{Sj}, \qquad Z_i \in M, \tag{127}$$

where

$$T_i = E\{\text{sojourn time in } Z_i\} = \int_0^\infty (1-Q_i(x))dx, \tag{128}$$

and p_{ij} are as in equation (83), i.e. given by $p_{ij} = Q_{ij}(\infty)$.

3.4.3 Stationary semi-Markov processes

To investigate the stationary state, let us assume that the embedded Markov chain is *irreducible* (every state can be reached from every other state with probability >0). In this case, the semi-Markov process is stationary if the initial distribution functions (eq.(121) are given by

$$A_{ij}(x) = \frac{p_i p_{ij}}{\sum_{k=0}^{m} p_k T_k} \int_0^x (1-F_{ij}(y))dy, \tag{129}$$

where $p_0, p_1, \dots, p_m$ is the *stationary distribution of the embedded Markov chain*, i.e.

the solution of

$$p_j = \sum_{k=0}^{m} p_{kj} p_k \quad \text{with } p_{jj} = 0 \quad \text{and} \quad \sum_{j=0}^{m} p_j = 1. \tag{130}$$

For a stationary semi-Markov process, the state probabilities (eq.(78)) are given by

$$P_i(t) = p_i T_i / \sum_{k=0}^{m} p_k T_k, \qquad i = 0,1,\ldots,m \tag{131}$$

for all $t \geq 0$. Furthermore, successive occurrence points of state Z_i constitute a stationary renewal process with renewal density

$$h_i(t) = p_i / \sum_{k=0}^{m} p_k T_k, \qquad i = 0,1,\ldots,m \tag{132}$$

for all $t \geq 0$.

If the semi-Markov process is not stationary but still has an irreducible Markov chain, then the conditional state probabilities (eq.(102)) and the state probabilities (eq.(78)) converge, for $t \to \infty$, to the value given by equation (131), independent of the initial conditions at $t = 0$.

3.5 Regenerative stochastic processes

A stochastic process with state space $Z_0, Z_1, \ldots, Z_m$ is said to be regenerative with respect to the state Z_k if successive transitions in Z_k (i.e. successive occurrence points of Z_k) constitute a renewal process - the so called *embedded renewal process*. For such a regenerative process, the state Z_k is a *regeneration state* and possesses the fundamental property that its occurrence is a regeneration point for the whole process, i.e. it brings the process to a situation of total independence from the preceding development. Between successive regeneration points, the stochastic behaviour of the process is a probabilistic replica of the behaviour between the first and the second regeneration point. Semi-Markov processes with a finite state space and an irreducible Markov chain are regenerative with respect to all states. Time-homogeneous Markov processes with a finite state space and an irreducible Markov chain are regenerative even at each time point. To be regenerative a process must contain at least one regeneration state.

If the process contains k regeneration states, say $Z_0, Z_1, \ldots, Z_{k-1}$, then this subset of the state space is a *semi-Markov process embedded in the regenerative process*, and the results obtained for semi-Markov processes can be applied to this subset. Because of the presence of non-regeneration states, a transition from a regeneration state Z_i to the same regeneration state through one or more non-regeneration states is allowed ($p_{ii} \geq 0$ instead of $p_{ii} = 0$ as by Markov and semi-Markov processes). Examples of this situation are shown later by equations (216) and (287).

If conditions (39) to (41) are satisfied by all embedded renewal processes, then the limits for $t \to \infty$ of the state probabilities $P_j(t) = Pr\{\xi(t) = Z_j\}$ exist, independently of the initial conditions at $t = 0$ [19,53-55], i.e.

$$\lim_{t \to \infty} P_j(t) = p_j \tag{133}$$

exist with $p_j \geq 0$ and

$$\sum_{j=0}^{m} p_j = 1. \tag{134}$$

The evaluation of p_j can often be performed using the final-value theorem of Laplace transforms.

3.6 Non-regenerative stochastic processes

The assumption of arbitrary (i.e. non-exponentially distributed) failure-free and repair times leads to non-regenerative stochastic processes, for systems with more than one element in the reliability block diagram. To these belong in particular superimposed renewal and alternating renewal processes, and some kinds of processes with a hierarchical structure. The investigation of such processes is not as simple as that for regenerative processes.

Besides problem-oriented investigation methods, a general approach consists of the transformation of the given stochastic process into a Markov process by a suitable choice of the state space. Three possibilities are known:

1. *Approximation of the distribution functions* [14,347]: A suitable approximation of the distribution functions of the failure-free and/or repair times can reduce the given process to a Markov process. In general the approximation is made with an Erlang distribution, which leads to the introduction of supplementary states (see Fig.19 later).
2. *Inclusion of supplementary variables* [13]: The given process becomes Markovian by introducing a sufficient number of supplementary variables. In reliability problems, these variables are the times elapsed from the last failure or from the last repair of a certain number of elements in the reliability block diagram. The method has been widely used, e.g. [98,112-114,118,121,131,158,191,192,199,228,232,287]. However, its practical utility is often limited by the complexity of the partial differential equations which are involved.

3. *Use of particular failure and repair rates* [71,185,229]: With the introduction for each element in the reliability block diagram of what we shall call Markovian failure and repair rates, defined as

$$\lambda^{\diamond}(t) = \lim_{\delta t \to 0} \frac{1}{\delta t} \Pr\{\text{failure in } (t,t+\delta t) \mid \text{up at } t\}, \quad \delta t > 0 \tag{135}$$

$$\mu^{\diamond}(t) = \lim_{\delta t \to 0} \frac{1}{\delta t} \Pr\{\text{repair in } (t,t+\delta t) \mid \text{down at } t\}, \quad \delta t > 0, \tag{136}$$

a Markov approach becomes possible. Based on equations (160) and (163) in section 4.4 it follows that

$$\lambda^{\diamond}(t) = \frac{p h_{udu}(t) + (1-p) h_{udd}(t)}{PA(t)} \tag{137}$$

$$\mu^{\diamond}(t) = \frac{p h_{duu}(t) + (1-p) h_{dud}(t)}{1-PA(t)} . \tag{138}$$

In the stationary state (eq.(72)) one obtains for all $t \geqslant 0$

$$\lambda^{\diamond}(t) = \frac{1}{MTTF} \tag{139}$$

$$\mu^{\diamond}(t) = \frac{1}{MTTR} , \tag{140}$$

with MTTF and MTTR as in equation (68). The method implies that the stochastic behaviour of each element is described by its own alternating renewal process; thus, it presents the same limitations as the methods based on Boolean representations [71].

The problem becomes easier if the reliability block diagram contains a large number of elements, because in this case the flow of failures can be approximated by a Poisson process [22,28,36,162].

CHAPTER 4

APPLICATIONS TO ONE-ITEM REPAIRABLE STRUCTURES

A one-item repairable structure is an entity of arbitrary complexity, which for investigative purposes is considered as a unit, and characterized by a distribution function of the failure-free times and one or more distribution functions for the repair and preventive maintenance times. Its reliability block diagram consists of a single element, Fig.15. Such a structure has been widely investigated in the lite-

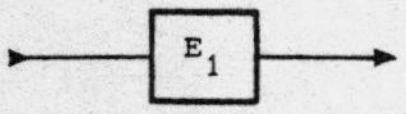

***Fig. 15.* Reliability block diagram for a one-item structure**

rature [68-85]. This chapter presents a summary of these efforts, and generalizes some results by assuming arbitrary initial conditions at $t = 0$ and by carefully investigating the asymptotic and stationary behaviours. The following assumptions hold:

1. The item alternates continuously from the operating state (up) to the repair state (down) and vice-versa.
2. Preventive maintenance is not considered.
3. After each repair the item is good-as-new.
4. Switching effects are negligible.
5. Failure-free and repair times (τ_i and τ_i', $i \geqslant 1$) are statistically independent and distributed according to $F(x)$ and $G(x)$, respectively.
6. At $t = 0$ the item is up with probability p or down with probability 1-p; the distribution function of the failure-free time starting at $t = 0$ (τ_0) is $F_A(x)$, that of the repair time starting at $t = 0$ (τ_0') is $G_A(x)$.
7. $F(0) = F_A(0) = G(0) = G_A(0) = 0$, the densitites $f_A(x)$, $f(x)$, $g_A(x)$ and $g(x)$ exist for all $x > 0$ and go to 0 as $x \to \infty$, $E\{\tau_i\} = MTTF$ and $E\{\tau_i'\} = MTTR$ as well as $Var\{\tau_i\}$ and $Var\{\tau_i'\}$ are $< \infty$ $(i \geqslant 1)$.

Assumption 1 is not restrictive if one considers operating time instead of calendar time, and if the concept *bad-as-old* can be used in the case of interruptions without repair. Assumption 3 holds if the item is completely renewed at each repair. Its validity must be checked carefully, however, if the unrenewed parts have not a (nearly) constant failure rate. Assumption 5 refines assumptions 3 and 4. Preventive maintenance and imperfect switching will be considered in chapter 6.

The stochastic behaviour of a repairable item which satisfies assumptions 1 through 7 above can be investigated using the alternating renewal process introduced in section 3.2 (Fig.12).

4.1 Reliability function

The reliability function R(t) is defined as

$$R(t) = Pr\{\text{item up in } (0,t) \mid \text{item up at } t=0\}, \tag{141}$$

and given by

$$R(t) = Pr\{\tau_0 > t\} = 1-F_A(t). \tag{142}$$

Equation (142) is similar to equation (1), because for a one-item structure, the failure-free time starting at $t=0$ (τ_0) is entirely characterized by the distribution function $F_A(x)$. From this and from equation (11) it follows for the mean time-to-failure MTTF that

$$MTTF = \int_0^{\infty} (1-F_A(t))dt. \tag{143}$$

4.2 Point-availability

The point-availability PA(t) is the probability of finding the item up at a time t

$$PA(t) = Pr\{\text{item up at } t\}. \tag{144}$$

Considering that at $t=0$ the item is up with probability p and down with probability 1-p, it follows for PA(t)

$$PA(t) = p\,Pr\{\text{item up at } t \mid \text{item up at } t=0\} + (1-p)Pr\{\text{item up at } t \mid \text{item down at } t=0\}.$$

The conditional probabilities involved in PA(t) can be evaluated with the same procedure used to develop equation (56). This leads to

$$PA(t) = p[1-F_A(t)+\int_0^t h_{duu}(x)(1-F(t-x))dx] + (1-p)\int_0^t h_{dud}(x)(1-F(t-x))dx. \tag{145}$$

From equations (145) and (70) follows then for the Laplace transform (eq.(28)) of PA(t):

$$\tilde{PA}(s) = p\,\frac{1-\tilde{f}_A(s)-[\tilde{f}(s)-\tilde{f}_A(s)]\tilde{g}(s)}{s[1-\tilde{f}(s)\tilde{g}(s)]} + (1-p)\,\frac{[1-\tilde{f}(s)]\tilde{g}_A(s)}{s[1-\tilde{f}(s)\tilde{g}(s)]}. \qquad (146)$$

Important in practical applications is the case of constant failure rate (λ), i.e. for $f_A x) = f(x) = \lambda e^{-\lambda x}$, which leads to

$$\tilde{PA}(s) = p\,\frac{1}{s+\lambda(1-\tilde{g}(s))} + (1-p)\,\frac{\tilde{g}_A(s)}{s+\lambda(1-\tilde{g}(s))}, \qquad (147)$$

and the case of constant failure and repair rates (λ and μ), i.e. for $f_A(x) = f(x) = \lambda e^{-\lambda x}$ and $g_A(x) = g(x) = \mu e^{-\mu x}$, which leads to

$$\tilde{PA}(s) = p\,\frac{s+\mu/p}{s(s+\lambda+\mu)}$$

and

$$PA(t) = \frac{\mu}{\lambda+\mu} + \frac{p(\lambda+\mu)-\mu}{\lambda+\mu}\,e^{-(\lambda+\mu)t}. \qquad (148)$$

Equation (145) can also be developed making a much deeper use of the regeneration properties of the alternating renewal process. Consider first the case $p = 1$ and $F_A(x) = F(x)$. For this case, the embedded renewal process (Fig.12)

$$\tau_0+\tau_1',\tau_1+\tau_2',\tau_2+\tau_3',\ldots$$

is ordinary, with renewal points $0, S_{duu1}, S_{duu2}, S_{duu3}, \ldots$. Let

$$PA_u(t) = \Pr\{\text{item up at } t \mid \text{item new at } t = 0\}. \qquad (149)$$

At the occurrence of S_{duu1}, i.e. when the first repair has just been finished, the item is new, as at $t = 0$. The embedded renewal process is independent at this time point of its previous development and finds itself in the same condition as at $t = 0$. From this follows that, for any $x \leqslant t$,

$$\Pr\{\text{item up at } t \mid S_{duu1} = x\} = PA_u(t-x). \qquad (150)$$

To evaluate $PA_u(t)$, consider now that the event

item up at t

occurs with one of the following mutually exclusive events

- no failure in $(0,t)$
- $S_{duu1} \leqslant t \cap$ item up at t.

For $PA_u(t)$ it follows then that

$$PA_u(t) = 1-F(t)+\int_0^t (f(x)*g(x))PA_u(t-x)dx. \tag{151}$$

The generalization to arbitrary initial conditions at $t=0$ leads finally to

$$PA(t) = p[1-F_A t)+\int_0^t (f_A(x)*g(x))PA_u(t-x)\,dx]$$
$$+(1-p)\int_0^t g_A(x)PA_u(t-x)dx. \tag{152}$$

The identity between equations (145) and (152) can be shown using Laplace transforms.

4.3 Interval-reliability

In many applications it is important to know the probability of having the item up not only at a particular time t, but also during an interval $(t,t+\theta)$. This probability is called interval-reliability $IR(t,t+\theta)$ and defined as

$$IR(t,t+\theta) = Pr\{\text{item up in } (t,t+\theta)\}, \quad \theta \geqslant 0. \tag{153}$$

For $\theta=0$ one has $IR(t,t+0) = PA(t)$. The evaluation of interval-reliability is similar to that of point-availability and leads to

$$IR(t,t+\theta) = p[1-F_A(t+\theta)+\int_0^t h_{duu}(x)(1-F(t+\theta-x))dx]$$
$$+(1-p)\int_0^t h_{dud}(x)(1-F(t+\theta-x))dx. \tag{154}$$

For the case of constant failure rate (λ), i.e. for $f_A(x) = f(x) = \lambda e^{-\lambda x}$, equation (154) becomes

$$IR(t,t+\theta) = PA(t)e^{-\lambda\theta}. \tag{155}$$

Relation (155), expressing the probability of being up in the interval $(t,t+\theta)$ as the product (probability of being up at t)·(probability of no failure in the subsequent interval of length θ), is a consequence of the *memoryless* property associated with the exponential distribution (eq.(66)).

4.4 Mission-oriented availabilities

Some other kinds of availabilities which are useful in practical applications are the average-availability, the joint-availability, the mission-availability and the work-mission-availability.

4.4.1 Average-availability

The average-availability AA(t) is defined as

$$AA(t) = \frac{1}{t}\, E\{\text{up time in } (0,t)\}, \qquad t>0. \tag{156}$$

AA(t) gives the expected fraction of the time interval (0,t) spent in the operating state. It is not difficult to see that the average-availability is related to the point-availability (eq.(145)) by the expression

$$AA(t) = \frac{1}{t} \int_0^t PA(x)\,dx. \tag{157}$$

Assuming constant failure and repair rates (λ and μ), i.e. for $f_A(x) = f(x) = \lambda e^{-\lambda x}$ and $g_A(x) = g(x) = \mu e^{-\mu x}$, equations (148) and (157) lead to

$$AA(t) = \frac{\mu}{\lambda+\mu} + \frac{p(\lambda+\mu)-\mu}{t(\lambda+\mu)^2}\,(1-e^{-(\lambda+\mu)t}). \tag{158}$$

4.4.2 Joint-availability

The joint-availability $JA(t, t+\theta)$ is defined as

$$JA(t,t+\theta) = Pr\{\text{item up at } t \cap \text{item up at } t+\theta\}, \qquad \theta \geqslant 0. \tag{159}$$

To evaluate $JA(t,t+\theta)$, let us consider that the event

item up at t ∩ item up at t+θ

occurs with one of the following mutually exclusive events

- item up in (t,t+θ)
- item up at t ∩ next failure occurs before t+θ ∩ item up at t+θ.

The probability of the first event is the interval-reliability (eq.(154)). For the second event, we must consider the distribution function of the *forward recurrence-*

time in the up state, $\tau_{Ru}(t)$ in Fig.16. From Fig.16 it follows that $\tau_{Ru}(t)$ can be

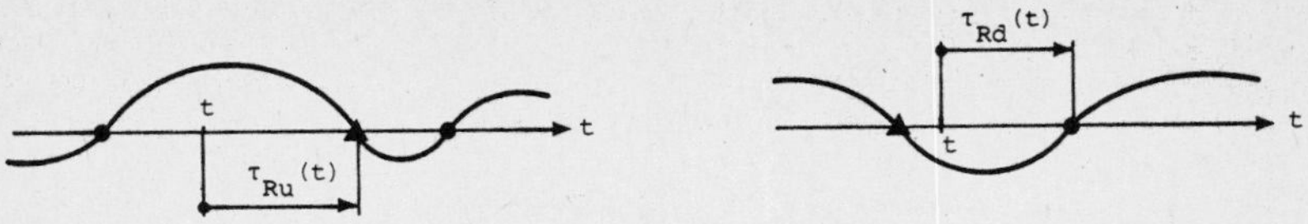

Fig. 16. **Forward recurrence-times in the up state ($\tau_{Ru}(t)$) and in the down state ($\tau_{Rd}(t)$) for a one-item repairable structure**

defined only if the item is up at the time t. This leads to

$$\Pr\{\tau_{Ru}(t) > x\} = \Pr\{\text{item up in } (t,t+x) \mid \text{item up at } t\}$$
$$= \frac{\Pr\{\text{item up in } (t,t+x)\}}{\Pr\{\text{item up at } t\}}$$

and then

$$\Pr\{\tau_{Ru}(t) \leqslant x\} = 1 - \frac{\Pr\{\text{item up in } (t,t+x)\}}{\Pr\{\text{item up at } t\}} = 1 - \frac{IR(t,t+x)}{PA(t)}, \tag{160}$$

where $PA(t)$ is the point-availability (eq.(145)) and $IR(t,t+x)$ is the interval-reliability (eq.(154)). Using the concept of the forward recurrence-time one finally obtains

$$JA(t,t+\theta) = IR(t,t+\theta) - \int_0^\theta \frac{\partial IR(t,t+x)}{\partial x} PA_d(\theta - x)\,dx, \tag{161}$$

where

$$PA_d(t) = \Pr\{\text{item up at } t \mid \text{item enters the down state at } t = 0\}$$

is given by equation (145) with $p = 0$ and $G_A(x) = G(x)$. For the case of constant failure rate (λ), i.e. for $f_A(x) = f(x) = \lambda e^{-\lambda x}$, equation (161) leads to

$$JA(t,t+\theta) = PA(t)PA_u(\theta), \tag{162}$$

where $PA_u(\theta)$ is obtained by equation (145) with $p = 1$ and $F_A(x) = F(x) = 1-e^{-\lambda x}$, or by equation (151) with $F(x) = 1-e^{-\lambda x}$. The investigation of the *forward recurrence-time in the down state,* $\tau_{Rd}(t)$ in Fig.16, leads to

$$\Pr\{\tau_{Rd}(t) \leqslant x\} = 1 - \frac{\Pr\{\text{item down in } (t,t+x)\}}{\Pr\{\text{item down at } t\}} \tag{163}$$

where $\Pr\{\text{item down at } t\} = 1 - PA(t)$ and

$$\Pr\{\text{item down in } (t,t+x)\} = p\int_0^t h_{udu}(y)(1-G(t+x-y))dy$$
$$+(1-p)[1-G_A(t+x)+\int_0^t h_{udd}(y)(1-G(t+x-y))dy]. \quad (164)$$

Equations (160) and (163) can be used to evaluate the Markovian failure and repair rates defined by equations (135) and (136) ($\lambda^\diamond(t) = \frac{\partial}{\partial x}\Pr\{\tau_{Ru}(t) \leqslant x\}$ and $\mu^\diamond(t) = \frac{\partial}{\partial x}\Pr\{\tau_{Rd}(t) \leqslant x\}$ at $x = 0$).

4.4.3 Mission-availability

The mission-availability $MA(T_O,t_f)$ is defined as

$MA(T_O,t_f)$ = Pr{each individual failure which occurs in a mission with total operating time T_O can be repaired in a time $\leqslant t_f$}. (165)

$MA(T_O,t_f)$ applies to the situations in which down times shorter than t_f can be accepted. From the definition of mission-availability, the item is up at the time $t = 0$ ($p = 1$). Considering that the end of the mission falls within an operating period, the expression for mission-availability is found by summing over all possibilities of having in the total operating time T_O exactly n failures ($n = 1,2,3,\ldots$), each of which can be repaired in a time shorter than t_f. This leads to

$$MA(T_O,t_f) = 1-F_A(T_O) + \sum_{n=1}^{\infty}[F_n(T_O)-F_{n+1}(T_O)](G(t_f))^n, \quad (166)$$

where $F_n(T_O)-F_{n+1}(T_O)$ is the probability of n failures when starting with τ_O distributed according to $F_A(x)$ (eq.(44)), and $(G(t_f))^n$ is the probability that each of the n repairs is shorter than t_f. Assuming a constant failure rate (λ), i.e. for $f_A(x) = f(x) = \lambda e^{-\lambda x}$, equation (166) becomes

$$MA(T_O,t_f) = e^{-\lambda T_O} + \sum_{n=1}^{\infty}\frac{(\lambda T_O)^n}{n!}e^{-\lambda T_O}(G(t_f))^n = e^{-\lambda T_O(1-G(t_f))}. \quad (167)$$

4.4.4 Work-mission-availability

The work-mission-availability $WMA(T_O,t_d)$ is defined as

$WMA(T_O,t_d)$ = Pr{the sum of all repair times for failures occurring in a mission with total operating time T_O is $\leqslant t_d$}. (168)

Similarly as with mission-availability (eq.(166)), one obtains (considering that one has $p = 1$)

$$WMA(T_O, t_d) = 1 - F_A(T_O) + \sum_{n=1}^{\infty} [F_n(T_O) - F_{n+1}(T_O)] GO_n(t_d), \tag{169}$$

where $GO_n(t_d)$ is the probability that the sum of n repair times, which are distributed according to G(x), is shorter than t_d. Setting $t = T_O + t_d$, WMA(t-x,x) is the distribution function of the total down time in the interval (0,t). From this it follows that

$$E\{\text{up time in } (0,t) \mid \text{item up at } t=0\} = \int_0^t WMA(t-x,x)\,dx. \tag{170}$$

The identity between equation (170) and the quantity $t \cdot AA(t)$ given by equation (157) with $p = 1$ can be shown using Laplace transforms. The function WMA(t-x,x) makes a jump equal to $1-F_A(t)$ at $x = 0$ and takes the value 1 at $x = t$. Its evaluation is difficult even in the case of constant failure and repair rates.

4.5 Asymptotic behaviour

As t becomes large, the point-availability (eq.(145)), the interval-reliability (eq.(154)), the average-availability (eq.(157)), the joint-availability (eq.(161)), and the distribution functions (eq.(160) and (163) of the forward recurrence-times $\tau_{Ru}(t)$ and $\tau_{Rd}(t)$ approach constant values which are independent of both the time origin and the initial conditions p, $F_A(x)$, and $G_A(x)$. Using the key renewal theorem (eq.(59)) and Laplace transforms for the joint-availability, one obtains

$$\lim_{t\to\infty} PA(t) = \frac{MTTF}{MTTF+MTTR} = PA \tag{171}$$

$$\lim_{t\to\infty} IR(t,t+\theta) = \frac{1}{MTTF+MTTR} \int_\theta^\infty (1-F(y))\,dy = PA \cdot [1 - \frac{1}{MTTF} \int_0^\theta (1-F(y))\,dy] = IR(\theta) \tag{172}$$

$$\lim_{t\to\infty} AA(t) = \frac{MTTF}{MTTF+MTTR} = AA = PA \tag{173}$$

$$\lim_{t\to\infty} JA(t,t+\theta) = \frac{MTTF}{MTTF+MTTR} PA_{us}(\theta) = JA(\theta) \tag{174}$$

$$\lim_{t\to\infty} Pr\{\tau_{Ru}(t) \leq x\} = \frac{1}{MTTF} \int_0^x (1-F(y))\,dy \tag{175}$$

$$\lim_{t\to\infty} Pr\{\tau_{Rd}(t) \leq x\} = \frac{1}{MTTR} \int_0^x (1-G(y))\,dy, \tag{176}$$

where $MTTF = E\{\tau_i\}$, $MTTR = E\{\tau'_i\}$ $(i \geq 1)$ and $PA_{us}(\theta)$ is the point-availability given by equation (145) with $p = 1$ and $F_A(x)$ as in equation (72).

4.6 Stationary state

As defined in section 3.2, an alternating renewal process is stationary if the conditions

$$p = \frac{MTTF}{MTTF+MTTR}, \quad F_A(x) = \frac{1}{MTTF}\int_0^x (1-F(y))dy, \quad G_A(x) = \frac{1}{MTTR}\int_0^x (1-G(y))dy \qquad (72)$$

hold. Taking these values for p, $F_A(x)$, and $G_A(x)$ in the equations (145),(154),(157),(161),(160) and (163), one obtains for the corresponding quantities the expressions derived for asymptotic behaviour (eq.(171) to (176)) for all $t \geq 0$. This correspondence between asymptotic behaviour and stationary state is useful in testing for *stationarity* in practical applications (see the interpretation given by equation (74)). The main results associated with one-item repairable structures in the stationary state are summarized in Table 8.

Quantity	Expression: Arbitrary failure and repair rates	Expression: Const. failure & repair rates (λ,μ)	Remarks, assumptions
1. Pr{item up at t = 0} (p)	$\frac{MTTF}{MTTF+MTTR}$	$\frac{\mu}{\lambda+\mu}$	$MTTF=E\{\tau_i\},\ i \geq 1$ $MTTR=E\{\tau'_i\},\ i \geq 1$
2. Distribution function of the up time starting at $t = 0$ $(F_A(x) = Pr\{\tau_0 \leq x\})$	$\frac{1}{MTTF}\int_0^x (1-F(y))dy$	$1-e^{-\lambda t}$	The same holds for the forward recurrence-time in the up state $(\tau_{Ru}(t))$
3. Distribution function of the down time starting at $t = 0$ $(G_A(x) = Pr\{\tau'_0 \leq x\})$	$\frac{1}{MTTR}\int_0^x (1-G(y))dy$	$1-e^{-\mu t}$	The same holds for the forward recurrence-time in the down state $(\tau_{Rd}(t))$
4. Renewal density $h_{du}(t)$, $h_{ud}(t)$	$\frac{1}{MTTF+MTTR}$	$\frac{\lambda\mu}{\lambda+\mu}$	$h_{du}(t) = ph_{duu}(t)+(1-p)h_{dud}(t)$ $h_{ud}(t) = ph_{udu}(t)+(1-p)h_{udd}(t)$ $p = MTTF/(MTTF+MTTR)$ as in point 1
5. Point-availability	$\frac{MTTF}{MTTF+MTTR}$	$\frac{\mu}{\lambda+\mu}$	$PA = Pr\{\text{item up at } t\},\ t \geq 0$
6. Average-availability	$\frac{MTTF}{MTTF+MTTR}$	$\frac{\mu}{\lambda+\mu}$	$AA = \frac{1}{t}E\{\text{up time in } (0,t)\},\ t>0$
7. Interval-reliability	$\frac{1}{MTTF+MTTR}\int_\theta^\infty (1-F(y))dy$	$\frac{\mu}{\lambda+\mu}e^{-\lambda\theta}$	$IR(\theta) = Pr\{\text{item up in } (t,t+\theta)\},\ t,\theta \geq 0$
8. Joint-availability	$\frac{MTTF}{MTTF+MTTR}PA_{us}(\theta)$	$(\frac{\mu}{\lambda+\mu})^2 + \frac{\lambda\mu e^{-(\lambda+\mu)\theta}}{(\lambda+\mu)^2}$	$JA = Pr\{\text{item up at } t \cap \text{item up at } t+\theta\}, \theta \geq 0$ $PA_{us}(\theta)=1-F_A(\theta)+\int_0^\theta h_{duu}(y)(1-F(\theta-y))dy$ $F_A(x)$ as in point 2; $h_{duu}=f_A*g+f_A*g*f*g+\ldots$

Table 8. **Main results for one-item repairable structures in a stationary state**

CHAPTER 5

APPLICATIONS TO SERIES, PARALLEL, AND SERIES/PARALLEL REPAIRABLE STRUCTURES

Series, parallel, and series/parallel combinations of elements are the basic structures of system reliability block diagrams (Tab.2 on p.6). In this chapter, these structures will be investigated under the following general assumptions:

1. The system alternates continuously from the operating state (up) to the repair state (down) and vice-versa.
2. Preventive maintenance is not considered.
3. No further failure can occur at system down.
4. The system has only one repair crew and repair is started without delay.
5. Redundant elements are repaired on-line.
6. After each repair, the repaired element is good-as-new.
7. Switching effects are negligible.
8. Failure-free and repair times are >0, continuous, statistically independent, and have finite mean and finite variance.

Assumption 1 has been discussed in chapter 4 (p.41). Assumption 3 has no influence on the reliability function but simplifies the availability investigations. Assumption 4 is realistic in many practical applications. Assumption 6 only refers to the element of the reliability block diagram which has been repaired (see the discussion of assumption 3 on p.41). Preventive maintenance and imperfect switching will be considered in chapter 6. For the same reliability block diagram, the distribution functions of the repair and failure-free times are successively generalized, starting with the case of constant failure and repair rates.

5.1 Series structures

A series structure arises in the case of a system without redundancy. The corresponding reliability block diagram is shown in Fig.17. Each element in Fig.17 is characterized by the distribution functions $F_i(x)$ of its failure-free times and $G_i(x)$ of its repair times.

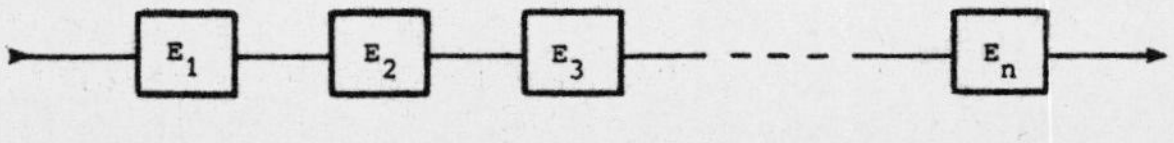

***Fig. 17.* Reliability block diagram for a system without redundancy (series structure)**

5.1.1 Constant failure and repair rates

Let us first assume that failure and repair rates are constant, i.e. that for i = 1,2, 3,...,n

$$F_i(x) = 1-e^{-\lambda_{Oi}x} \tag{177}$$

and

$$G_i(x) = 1-e^{-\mu_{iO}x} \tag{178}$$

hold. Because of equations (177) and (178), the stochastic behaviour of the system is described by a *time-homogeneous Markov process*. Let Z_O be the system up state and Z_i the state in which element i is down. The corresponding transition probabilities diagram for an arbitrary time interval $(t,t+\delta t)$ is given in Fig.18. From Fig.18, and

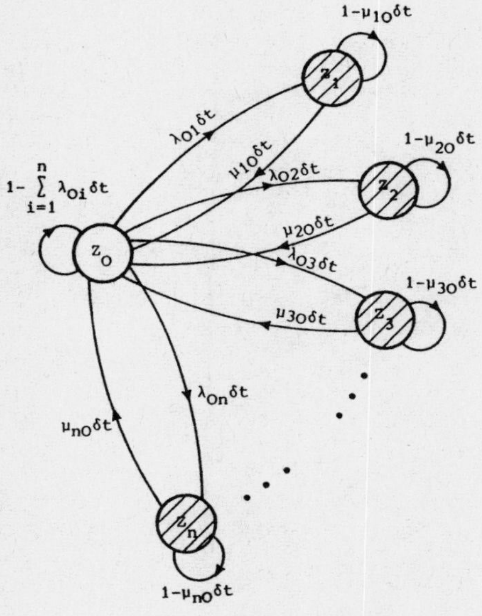

Fig. 18. **Diagram of the transition probabilities in $(t,t+\delta t)$, and transition rates (ρ_{ij}), for a repairable series structure (constant failure and repair rates, only one repair crew, no further failure at system down, t arbitrary, $\delta t \to 0$)**

taking care of equation (100), one obtains the following system of differential equations which govern stochastic behaviour of the series structure:

$$\dot{P}_O(t) = -\lambda_O P_O(t) + \sum_{i=1}^{n} \mu_{iO} P_i(t)$$

$$\dot{P}_i(t) = -\mu_{iO} P_i(t) + \lambda_{Oi} P_O(t), \qquad i = 1,2,\ldots,n \tag{179}$$

with

$$\lambda_O = \sum_{i=1}^{n} \lambda_{Oi}. \tag{180}$$

(λ_O corresponds to λ_S in eq.(16).) The reliability function

$$R_{SO}(t) = \Pr\{\text{system up in } (0,t) \mid Z_O \text{ is entered at } t=0\}, \tag{181}$$

is given by (eq.(15), or (111), or (107) to (110))

$$R_{SO}(t) = e^{-\lambda_O t}, \tag{182}$$

with λ_O as in equation (180). From equations (182) and (27), it then follows for the mean time-to-system-failure $MTTF_{SO}$ that

$$MTTF_{SO} = 1/\lambda_O. \tag{183}$$

Assuming that the system enters state Z_O at $t=0$, the point-availability

$$PA_{SO}(t) = \Pr\{\text{system up at } t \mid Z_O \text{ is entered at } t=0\}, \tag{184}$$

is obtained by solving equations (122) and (105), or equation (179) with the initial conditions $P_O(0) = 1$ and $P_i(0) = 0$ for $i = 1,2,\ldots,n$ and with $PA_{SO}(t) = P_O(t)$. For the Laplace transform (eq.(28)) of $PA_{SO}(t)$ follows

$$\tilde{PA}_{SO}(s) = \frac{1}{s\left(1+\sum_{i=1}^{n} \frac{\lambda_{Oi}}{s+\mu_{iO}}\right)}. \tag{185}$$

Assuming again that the system enters state Z_O at $t=0$, and considering equations (177), (155), and (182), one obtains for the interval-reliability

$$IR_{SO}(t,t+\theta) = \Pr\{\text{system up in } (t,t+\theta) \mid Z_O \text{ is entered at } t=0\}, \quad \theta \geqslant 0, \tag{186}$$

the expression

$$IR_{SO}(t,t+\theta) = PA_{SO}(t)\, e^{-\lambda_O \theta}, \tag{187}$$

with λ_O as in equation (180).

Asymptotic behaviour exists, independent of the initial conditions at $t=0$, and leads to

$$\lim_{t\to\infty} PA_{SO}(t) = \lim_{s\to 0} s\,\tilde{PA}_{SO}(s) = PA_S = \frac{1}{1+\sum_{i=1}^{n}\frac{\lambda_{Oi}}{\mu_{iO}}} \,. \tag{188}$$

Expression (188) also gives stationary-state point-availability and average-availability (eq. (156)). For interval-reliability, equation (187) holds with $PA_{SO}(t) = PA_S$ as in equation (188).

5.1.2 Constant failure rates and arbitrary repair rates

Generalization of the repair times distribution functions ($G_i(x)$ with density $g_i(x)$) leads to a *semi-Markov process*. Considering equation (177) and using Fig.18 only to visualize the *state transition diagram,* it follows for the semi-Markov transition probabilities $Q_{ij}(x)$ (eq.(87))

$$Q_{Oi}(x) = \frac{\lambda_{Oi}}{\lambda_O}\,(1-e^{-\lambda_O x})$$

$$Q_{iO}(x) = G_i(x), \qquad i = 1,2,3,\ldots,n, \tag{189}$$

with λ_O as in equation (180).

The reliability function is still given by equation (182). For the point-availability (eq.(184)) one obtains from equation (122)

$$P_{OO}(t) = e^{-\lambda_O t} + \sum_{i=1}^{n}\int_0^t \lambda_{Oi}e^{-\lambda_O x}\,P_{iO}(t-x)\,dx,$$

$$P_{iO}(t) = \int_0^t g_i(x)\,P_{OO}(t-x)\,dx, \qquad i = 1,2,3,\ldots,n, \tag{190}$$

and from equation (105)

$$PA_{SO}(t) = P_{OO}(t). \tag{191}$$

The Laplace transform (eq.(28)) of $PA_{SO}(t)$ is then given by

$$\tilde{PA}_{SO}(s) = \frac{1}{s+\sum_{i=1}^{n}\lambda_{Oi}(1-\tilde{g}(s))} \,. \tag{192}$$

The interval-reliability (eq.(186)) is obtained using equation (187) with $PA_{SO}(t)$ from equation (192).

Asymptotic behaviour exists, independent of the initial conditions at $t = 0$, and leads to

$$\lim_{t\to\infty} PA_{SO}(t) = PA_S = \frac{1}{1+\sum_{i=1}^{n}\lambda_{Oi}MTTR_i}, \tag{193}$$

with

$$MTTR_i = E\{\text{repair time of element } E_i\} = \int_O^\infty (1-G_i(x))dx. \tag{194}$$

Expression (193) also gives stationary-state point-availability and average-availability (eq.(156)). For the interval-reliability, equation (187) holds with $PA_{SO}(t) = PA_S$ as in equation (193).

5.1.3 Arbitrary failure and repair rates

Generalization of repair and failure-free times distribution functions leads to a *non-regenerative stochastic process*. This model can be investigated using supplementary variables, or by approximating the distribution functions of the failure-free times in such a way that the involved stochastic process is reduced to a regenerative process. Using for approximation an Erlang distribution function, the process is semi-Markovian.

As an example, let us consider the case of a two-element series structure and assume that the repair times are arbitrary, with densities $g_{1O}(x)$ and $g_{2O}(x)$, and that the failure-free times have densities

$$f_1(x) = \lambda_{O1}^2 x e^{-\lambda_{O1}x} \tag{195}$$
$$f_2(x) = \lambda_{O2} e^{-\lambda_{O2}x}. \tag{196}$$

Equation (195) is the density of the sum of two exponentially distributed random time intervals with density $\lambda_{O1}e^{-\lambda_{O1}x}$. Under these assumptions, the two-element series structure corresponds to a 1-out-of-2 standby redundancy, with constant failure rate λ_{O1}, in series with an element with constant failure rate λ_{O2}. Fig.19 gives the equivalent reliability block diagram and the corresponding *state transition diagram*. Z_O is the system up state, and $Z_{1'}, Z_{2'}$ are supplementary states, necessary only for computation.

For the semi-Markov transition probabilities $Q_{ij}(x)$, one obtains from equation (87) and Fig.19

$$Q_{O1'}(x) = Q_{1'1}(x) = \frac{\lambda_{O1}}{\lambda_{O1}+\lambda_{O2}}\,(1-e^{-(\lambda_{O1}+\lambda_{O2})x})$$

$$Q_{O2}(x) = Q_{1'2'}(x) = \frac{\lambda_{O2}}{\lambda_{O1}+\lambda_{O2}}\,(1-e^{-(\lambda_{O1}+\lambda_{O2})x})$$

$$Q_{20}(x) = Q_{2'1'}(x) = \int_0^x g_{20}(y)\,dy$$

$$Q_{10}(x) = \int_0^x g_{10}(y)\,dy \; . \tag{197}$$

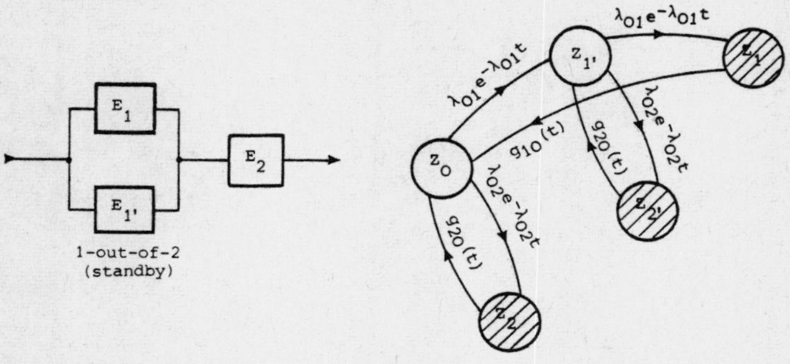

Fig. 19. **Equivalent reliability block diagram and corresponding state transition diagram for a two-element series repairable structure with Erlangian-distributed failure-free times for element E_1 (arbitrary repair rates, only one repair crew, no further failure at system down)**

The reliability function (eq.(181)) is given by (eq.(15) and (32), or (126))

$$R_{S0}(t) = (1+\lambda_{01}t)e^{-\lambda_{01}t}e^{-\lambda_{02}t} = (1+\lambda_{01}t)e^{-(\lambda_{01}+\lambda_{02})t}, \tag{198}$$

and the mean time-to-system-failure by (eq.(27))

$$MTTF_{S0} = \frac{2\lambda_{01}+\lambda_{02}}{(\lambda_{01}+\lambda_{02})^2} \; . \tag{199}$$

For the point-availability $PA_{S0}(t)$ (eq.(184)), the solution of equations (122) and (105) with $Q_{ij}(x)$ as in equation (197) leads to the following Laplace transform (eq.(28)) of $PA_{S0}(t) = P_{00}(t)+P_{01'}(t)$

$$\tilde{PA}_{S0}(s) = \frac{s+2\lambda_{01}+\lambda_{02}(1-\tilde{g}_{20}(s))}{[s+\lambda_{01}+\lambda_{02}(1-\tilde{g}_{20}(s))]^2-\lambda_{01}^2\tilde{g}_{10}(s)} \; . \tag{200}$$

The interval-reliability (eq.(186)) can be computed from

$$IR_{S0}(t,t+\theta) = P_{00}(t)R_{S0}(\theta)+P_{01'}(t)R_{S1'}(\theta), \tag{201}$$

where the pairs $P_{00}(t)$, $P_{01'}(t)$ and $R_{S0}(\theta)$, $R_{S1'}(\theta)$ are obtained by solving equations (122) and (126) respectively, with $Q_{ij}(x)$ as in equation (197).

Asymptotic behaviour exists, independent of the initial conditions at $t=0$, and leads to

$$\lim_{t\to\infty} PA_{SO}(t) = PA_S = \frac{2}{2+2\lambda_{O2}MTTR_2+\lambda_{O1}MTTR_1} \tag{202}$$

and

$$\lim_{t\to\infty} IR_{SO}(t,t+\theta) = IR_S(\theta) = \frac{(2+\lambda_{O1}\theta)e^{-(\lambda_{O1}+\lambda_{O2})\theta}}{2+2\lambda_{O2}MTTR_2+\lambda_{O1}MTTR_1}, \tag{203}$$

with $MTTR_i$ as in equation (194). Expression (202) also gives stationary-state point-availability and average-availability (eq.(156)). This remark holds also for equation (203).

Table 9 summarizes the main results obtained for series structures.

Quantity	Expression	Remarks, assumptions
1. Reliability function ($R_{SO}(t)$)	$\prod_{i=1}^{n} R_i(t)$	Elements $E_1, E_2, \ldots, E_n$ are independent
2. Mean time-to-system-failure ($MTTF_{SO}$)	$\int_0^\infty R_{SO}(t)dt$	$R_i(t)=e^{-\lambda_{Oi}t} \rightarrow R_{SO}(t)=e^{-\lambda_O t}$, $MTTF_{SO}=1/\lambda_O$, $\lambda_O=\lambda_{O1}+\lambda_{O2}+\ldots+\lambda_{On}$
3. System failure rate ($\lambda_O(t)$)	$\sum_{i=1}^{n} \lambda_{Oi}(t)$	Elements $E_1, E_2, \ldots, E_n$ are independent
4. Stationary-state point-availability and average-availability ($PA_S = AA_S$)	1) $\frac{1}{1+\sum_{i=1}^{n}\frac{\lambda_{Oi}}{\mu_{iO}}}$ 2) $\frac{1}{1+\sum_{i=1}^{n}\lambda_{Oi}MTTR_i}$ 3) $\frac{2}{2+2\lambda_{O2}MTTR_2+\lambda_{O1}MTTR_1}$	At system down, no further failure can occur 1) Constant failure rate (λ_{Oi}) and constant repair rate (μ_{iO}) for each element 2) Constant failure rate (λ_{Oi}) for each element; $MTTR_i$ = mean time-to-repair of element E_i 3) 2-element series structure with failure rates $\lambda_{O1}^2 t/(1+\lambda_{O1}t)$ for E_1 and λ_{O2} for E_2
5. Stationary-state interval-reliability ($IR_S(\theta)$) in the case of constant failure rate	$PA_S \cdot e^{-\lambda_O\theta}$	Each element has a constant failure rate (λ_{Oi}), $\lambda_O = \lambda_{O1}+\lambda_{O2}+\ldots+\lambda_{On}$)

Table 9. Main results for repairable systems without redundancy (one repair crew, no further failure at system down, independent elements)

5.2 1-out-of-2 redundancies

The 1-out-of-2 redundancy is the simplest redundant structure encountered in practical applications. It consists of two elements E_1 and E_2, one of which is in the operating state and the other in reserve. When a failure occurs, one element is put in repair and the other continues the operation. Assuming ideal switching and failure detection, the reliability block diagram is a parallel connection of elements E_1 and E_2, Fig.20. Such a model has been widely investigated in the literature [86-157]. This section summarizes these efforts and extends some results.

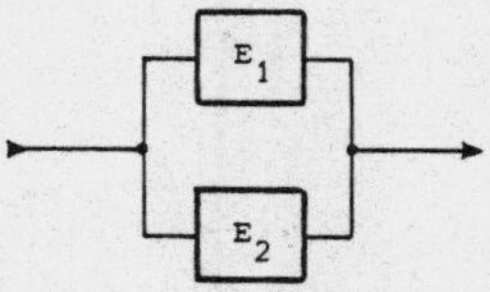

Fig. 20. **Reliability block diagram for 1-out-of-2 redundancy**

5.2.1 Constant failure and repair rates

Let us consider first the case of two identical elements in warm redundancy, with constant failure and repair rates, i.e. with distribution functions

$$F(x) = 1-e^{-\lambda x} \tag{204}$$

for the failure-free times in the operating state,

$$F_r(x) = 1-e^{-\lambda_r x} \tag{205}$$

for the failure-free times in the reserve state and

$$G(x) = 1-e^{-\mu x} \tag{206}$$

for the repair times. With these assumptions, the stochastic behaviour of the system is described by a *time-homogeneous Markov process*. The corresponding transition probabilities diagram for an arbitrary time interval $(t,t+\delta t)$ has been given in Fig.13. In state Z_i, i elements are down. The system has only one repair crew and the repair in state Z_1 does not affect the item in the operating state.

For the system of Fig.13, the reliability functions $R_{S0}(t)$ and $R_{S1}t)$ (eq.(106)) are obtained from equation (111) by solving

$$R_{S0}(t) = e^{-(\lambda+\lambda_r)t} + \int_0^t (\lambda+\lambda_r) e^{-(\lambda+\lambda_r)x} R_{S1}(t-x)\,dx$$

$$R_{S1}(t) = e^{-(\lambda+\mu)t} + \int_0^t \mu e^{-(\lambda+\mu)x} R_{S0}(t-x)\,dx \,. \tag{207}$$

In particular, the Laplace transform (eq.(28)) of $R_{S0}(t)$ and the mean time-to-system-failure $MTTF_{S0}$ are given by

$$\tilde{R}_{S0}(s) = \frac{s+2\lambda+\lambda_r+\mu}{(s+\lambda+\lambda_r)(s+\lambda)+s\mu} \tag{208}$$

and

$$MTTF_{S0} = \tilde{R}_{S0}(0) = \frac{2\lambda+\lambda_r+\mu}{\lambda(\lambda+\lambda_r)} \,. \tag{209}$$

The point-availability $PA_{S0}(t)$ (eq.(184)) is obtained by solving equations (122) and (105), or equation (99) with the initial conditions $P_0(0) = 1$ and $P_1(0) = P_2(0) = 0$, and with $PA_S(t) = P_0(t)+P_1(t)$. From equation (122) it follows that

$$P_{00}(t) = e^{-(\lambda+\lambda_r)t} + \int_0^t (\lambda+\lambda_r) e^{-(\lambda+\lambda_r)x} P_{10}(t-x)\,dx$$

$$P_{10}(t) = \int_0^t \mu e^{-(\lambda+\mu)x} P_{00}(t-x)\,dx + \int_0^t \lambda e^{-(\lambda+\mu)x} P_{20}(t-x)\,dx$$

$$P_{20}(t) = \int_0^t \mu e^{-\mu x} P_{10}(t-x)\,dx$$

$$P_{11}(t) = e^{-(\lambda+\mu)t} + \int_0^t \mu e^{-(\lambda+\mu)x} P_{01}(t-x)\,dx + \int_0^t \lambda e^{-(\lambda+\mu)x} P_{21}(t-x)\,dx$$

$$P_{01}(t) = \int_0^t (\lambda+\lambda_r) e^{-(\lambda+\lambda_r)x} P_{11}(t-x)\,dx$$

$$P_{21}(t) = \int_0^t \mu e^{-\mu x} P_{11}(t-x)\,dx \,, \tag{210}$$

and then (eq.(105))

$$PA_{S0}(t) = P_{00}(t)+P_{01}(t) \,. \tag{211}$$

The solution of equations (210) and (211) leads to the following Laplace transform (eq.(28)) of $PA_{S0}(t)$:

$$\tilde{PA}_{S0}(s) = \frac{(s+\mu)(s+\lambda+\lambda_r+\mu)+s\lambda}{s[(s+\lambda+\lambda_r)(s+\lambda+\mu)+\mu(s+\mu)]} \,. \tag{212}$$

Equation (210) can also be used to compute the point-availabilities $PA_{S1}(t)$ and $PA_{S2}(t)$, according to equations (104) and (105).

The interval-reliability $IR_{SO}(t,t+\theta)$ (eq.(186)) is given by

$$IR_{SO}(t,t+\theta) = P_{OO}(t)R_{SO}(\theta)+P_{O1}(t)R_{S1}(\theta), \tag{213}$$

where the pairs $P_{OO}(t)$, $P_{O1}(t)$ and $R_{SO}(\theta)$, $R_{S1}(\theta)$ are obtained by solving equations (210) and (207) respectively.

Asymptotic behaviour exists, independent of the initial conditions at $t=0$, and leads to

$$\lim_{t\to\infty} PA_{SO}(t) = PA_S = \frac{\mu(\lambda+\lambda_r+\mu)}{(\lambda+\lambda_r)(\lambda+\mu)+\mu^2} \tag{214}$$

$$\lim_{t\to\infty} IR_{SO}(t,t+\theta) = IR_S(\theta) = \frac{\mu^2 R_{SO}(\theta)+\mu(\lambda+\lambda_r)R_{S1}(\theta)}{(\lambda+\lambda_r)(\lambda+\mu)+\mu^2}\,. \tag{215}$$

Expression (214) also gives stationary-state point-availability and average-availability (eq.(156)). The same remark holds also for equation (215).

With $\lambda_r \equiv 0$ and $\lambda_r = \lambda$ one obtains the results for the *standby* and for the *active redundancy* cases, respectively. The influence of *load sharing* can be considered by modifying the failure rates (and thus the transition probabilities) at the corresponding state changes. The case of *different elements*, with failure rates λ_1 and λ_2, and repair rates μ_1 and μ_2, can be investigated using the transition probabilities diagram given in Fig.21. (For $\mu_1=\mu_2=\mu$, Fig.21 leads to Fig.14a on p.31.)

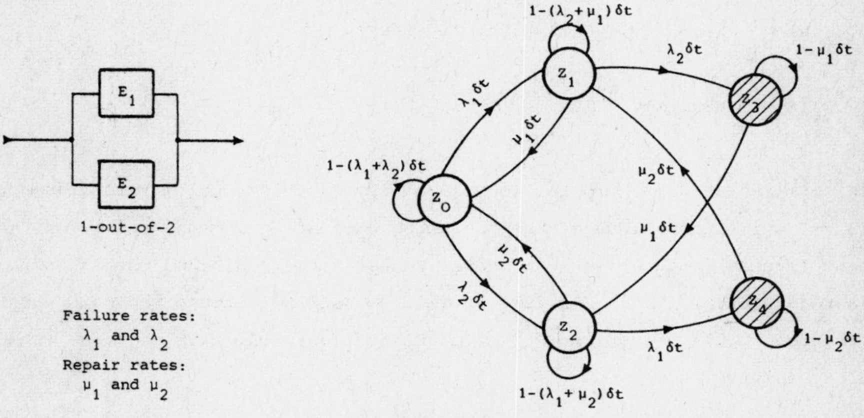

$\rho_{O1}=\lambda_1;\rho_{O2}=\lambda_2;\rho_{1O}=\mu_1;\rho_{2O}=\mu_2;\ \rho_{13}=\lambda_2;\rho_{24}=\lambda_1;\rho_{32}=\mu_1;\rho_{41}=\mu_2$

Fig. 21. **Diagram of the transition probabilities in $(t,t+\delta t)$, and transition rates (ρ_{ij}), for a repairable 1-out-of-2 redundancy with different elements (constant failure and repair rates, only one repair crew, t arbitrary, $\delta t \to 0$)**

5.2.2 Constant failure rates and arbitrary repair rate

Generalization of the repair rate leads to a *regenerative stochastic process with only two regeneration states,* states Z_0 and Z_1 in Fig.13. These two states constitute an *embedded semi-Markov process* on which the investigation can be based. Fig.22 gives a time schedule of the process. Assuming the failure rates λ and λ_r as in equa-

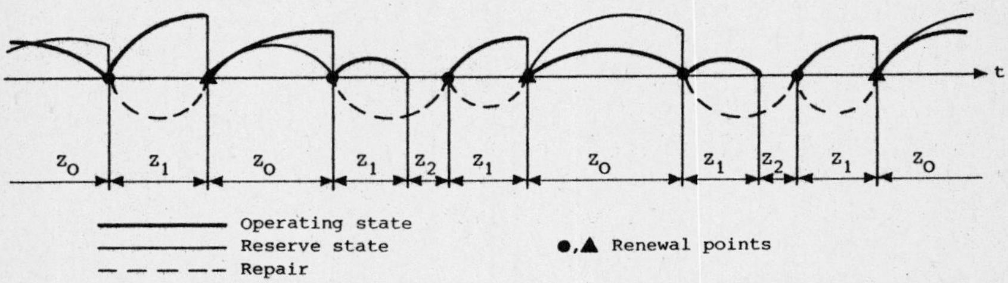

Fig. 22. **Time schedule of a repairable 1-out-of-2 warm redundancy (constant failure rates (λ,λ_r), arbitrary repair rate; for clarity, the repair times have been exaggerated)**

tions (204) and (205), and the repair times distributed according to G(x), with density g(x), one obtains for the embedded semi-Markov process the following semi-Markov transition probabilities $Q_{ij}(x)$ (eq.(87) and Fig.22):

$$Q_{01}(x) = 1-e^{-(\lambda+\lambda_r)x}$$

$$Q_{12}(x) = \int_0^x \lambda e^{-\lambda y}(1-G(y))dy = 1-e^{-\lambda x}-\int_0^x \lambda e^{-\lambda y}G(y)dy$$

$$Q_{10}(x) = \int_0^x g(y)e^{-\lambda y}dy = G(x)e^{-\lambda x}+\int_0^x \lambda e^{-\lambda y}G(y)dy$$

$$Q_{121}(x) = \int_0^x g(y)(1-e^{-\lambda y})dy. \qquad (216)$$

$Q_{12}(x)$ will be used to compute the reliability function. $Q_{121}(x)$ implies a transition $Z_1 \to Z_2 \to Z_1$, i.e. a failure of the operating element during the repair of the second element (system failure) with the consequent return to Z_1, and considers that Z_2 is *not* a regeneration state (Fig.22). It will be used to compute the point-availability.

The reliability functions $R_{S0}(t)$ and $R_{S1}(t)$ (eq.(106)) are obtained from equation (126) by solving

$$R_{S0}(t) = e^{-(\lambda+\lambda_r)t}+\int_0^t (\lambda+\lambda_r)e^{-(\lambda+\lambda_r)x}R_{S1}(t-x)dx$$

$$R_{S1}(t) = e^{-\lambda t}(1-G(t))+\int_0^t g(x)e^{-\lambda x}R_{S0}(t-x)dx\,. \qquad (217)$$

In particular, the Laplace transform (eq.(28)) of $R_{SO}(t)$ and the mean time-to-system-failure $MTTF_{SO}$ are given by

$$\tilde{R}_{SO}(s) = \frac{s+\lambda+(\lambda+\lambda_r)(1-\tilde{g}(s+\lambda))}{(s+\lambda)[s+(\lambda+\lambda_r)(1-\tilde{g}(s+\lambda))]} \tag{218}$$

and

$$MTTF_{SO} = \tilde{R}_{SO}(0) = \frac{1}{\lambda} + \frac{1}{(\lambda+\lambda_r)(1-\tilde{g}(\lambda))} . \tag{219}$$

The point-availabilities $PA_{SO}(t)$, $PA_{S1}(t)$ and $PA_{S2}(t)$ (eq.(104)) are obtained by solving equations (122) and (105) with $Q_{ij}(x)$ as in equation (216), i.e. by solving

$$\begin{aligned}
P_{OO}(t) &= e^{-(\lambda+\lambda_r)t}+\int_0^t (\lambda+\lambda_r)e^{-(\lambda+\lambda_r)x} P_{1O}(t-x)dx \\
P_{1O}(t) &= \int_0^t g(x)e^{-\lambda x} P_{OO}(t-x)dx+\int_0^t g(x)(1-e^{-\lambda x}) P_{1O}(t-x)dx \\
P_{11}(t) &= (1-G(t))e^{-\lambda t}+\int_0^t g(x)e^{-\lambda x} P_{O1}(t-x)dx+\int_0^t g(x)(1-e^{-\lambda x}) P_{11}(t-x)dx \\
P_{O1}(t) &= \int_0^t (\lambda+\lambda_r)e^{-(\lambda+\lambda_r)x} P_{11}(t-x)dx.
\end{aligned} \tag{220}$$

This leads in particular to the following Laplace transform (eq.(28)) of $PA_{SO}(t) = P_{OO}(t)+P_{O1}(t)$:

$$\tilde{PA}_{SO}(s) = \frac{(s+\lambda)(1-\tilde{g}(s))+\lambda_r(1-\tilde{g}(s+\lambda))+\lambda+s\tilde{g}(s+\lambda)}{(s+\lambda)[(s+\lambda+\lambda_r)(1-\tilde{g}(s))+s\tilde{g}(s+\lambda)]} . \tag{221}$$

The interval-reliability (eq.(186)) can be approximated using equation (213), with $P_{OO}(t)$ and $P_{O1}(t)$ from equation (220), and $R_{SO}(\theta)$ and $R_{S1}(\theta)$ from equation (217). The approximation, which assumes that state Z_1 is regenerative at each time point (Fig.22), is good if $MTTR \ll 1/\lambda$ holds, where MTTR is given by

$$MTTR = E\{\text{repair time}\} = \int_0^\infty (1-G(x))dx. \tag{222}$$

Asymptotic behaviours exists, independent of the initial conditions at $t=0$, and leads to

$$\lim_{t\to\infty} PA_{SO}(t) = PA_S = \frac{\lambda+\lambda_r(1-\tilde{g}(\lambda))}{\lambda(\lambda+\lambda_r)MTTR+\lambda\tilde{g}(\lambda)} , \tag{223}$$

with MTTR as in equation (222). Expression (223) also gives stationary-state point-availability and average-availability (eq.(156)).

5.2.3 Influence of the repair times density shape

To investigate the influence of the shape of the repair times density function on mean time-to-system-failure and on point-availability in the stationary state, let us assume that

$$g(x) = \begin{cases} 0 & \text{for } x < \psi \\ \mu' e^{-\mu'(x-\psi)} & \text{for } x \geq \psi . \end{cases} \quad (224)$$

g(x) as defined by equation (224) can be the approximation of a real density function of repair times, as shown in Fig.23. For the investigation let $\psi\lambda \ll 1$ and μ' such

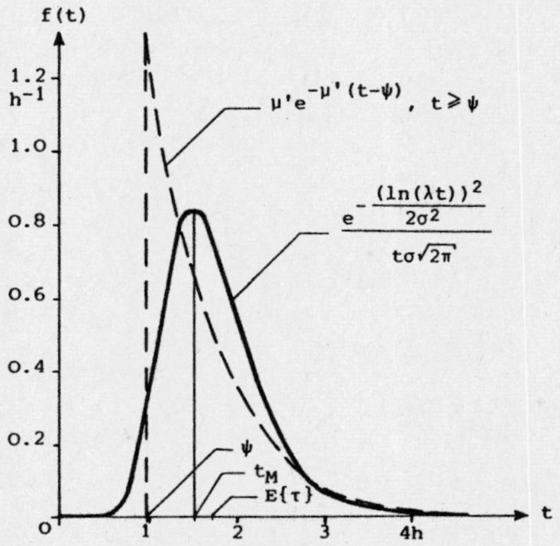

Fig. 23. **Approximation of a lognormal density by a shifted exponential density with the same mean**

that the mean time-to-repair is the same as with $g(x) = \mu e^{-\mu x}$, i.e. such that

$$\text{MTTR} = \psi + \frac{1}{\mu'} = \frac{1}{\mu} \quad (225)$$

holds. From

$$\tilde{g}(s) = \int_0^\infty g(t)\, e^{-st} dt = \frac{\mu'}{s+\mu'}\, e^{-\psi s}$$

follows then

$$\tilde{g}(\lambda) = \frac{\mu' e^{-\psi\lambda}}{\lambda+\mu'} \simeq \frac{\mu'(1-\psi\lambda)}{\lambda+\mu'} = \frac{\mu(1-\psi\lambda)}{\lambda+\mu(1-\psi\lambda)} . \quad (226)$$

Putting $\tilde{g}(\lambda)$ in equations (219) and (223) one obtains

$$\frac{\text{MTTF}_{SO} \text{ for } \psi > 0}{\text{MTTF}_{SO} \text{ for } \psi = 0} \simeq 1 - \lambda\psi \quad (227)$$

and

$$\frac{PA_{S \text{ for } \psi > 0}}{PA_{S \text{ for } \psi = 0}} \approx 1+(\lambda\psi)^2 \approx 1. \qquad (228)$$

Equations (227) and (228) show that as long as the mean time-to-repair (MTTR) is not changed and MTTR $\ll 1/\lambda$, the shape of the repair times density function has little influence on the mean time-to-system-failure or on the stationary value of the point-availability. The procedure introduced here can be applied to other reliability models. A further approach is to use a Taylor expansion of $\tilde{g}(\lambda)$ [9], or some other limit expressions [22-27,51,56-59]. Research continues in this field.

5.2.4 Constant failure rate in the reserve state, arbitrary failure rate in the operating state, and arbitrary repair rates

Generalization of operating- and reserve-state distribution functions for repair and failure-free times leads to a *non-regenerative stochastic process*. However, in many practical applications, the situation arise in which a constant reserve-state failure rate can be assumed. In this case the associated stochastic process is *regenerative with only one regeneration state*. The distribution functions involved are: $F(x)$ for the failure-free times in the operating state, $V(x) = 1-e^{-\lambda_r x}$ for the failure-free times in the reserve state, $G(x)$ for operating state failure repair times, and $W(x)$ for reserve state failure repair times, with densities $f(x)$, $\lambda_r e^{-\lambda_r x}$, $g(x)$ and $w(x)$ respectively. Fig.24 gives the corresponding *state transition*

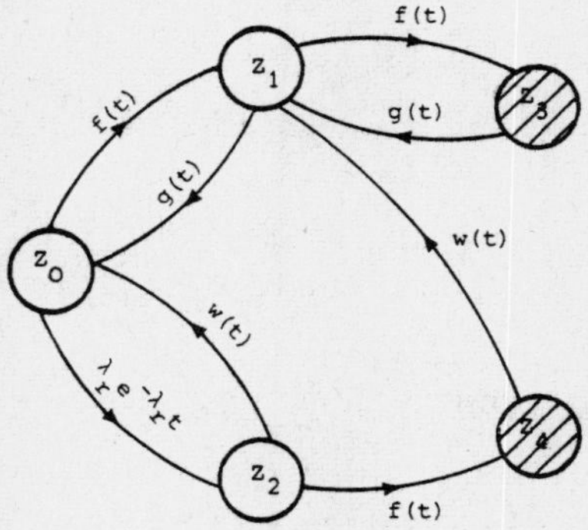

Fig. 24. **State transition diagram for a repairable 1-out-of-2 redundancy (failure rate arbitrary in the operating state and constant in the reserve state, arbitrary repair rates, only one repair crew)**

diagram. The system is down in states Z_3 and Z_4. State Z_1 is the *regeneration state*. Its occurrence is a regeneration point for the whole process (section 3.5). To give an idea of the stochastic behaviour of the system, Fig.25 shows a time schedule.

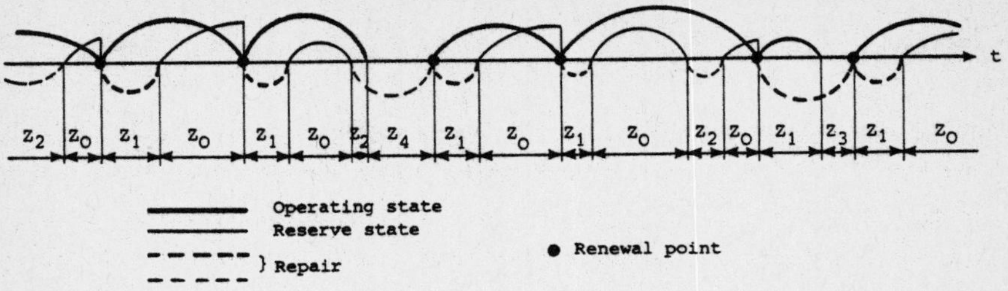

Fig. 25. **Time schedule of the 1-out-of-2 redundancy described by the model of Fig.24 (for clarity, the repair times have been exaggerated)**

5.2.4.1 At $t=0$ the system enters the regeneration state, Z_1

Let us consider first the case in which the system enters regeneration state Z_1 at $t=0$. We designate occurrence of the first renegeration point after $t=0$ as S_{RP1}.

The reliability function

$$R_{S1}(t) = \Pr\{\text{system up in } (0,t) \mid Z_1 \text{ is entered at } t=0\}, \tag{229}$$

is given by

$$R_{S1}(t) = 1-F(t)+\int_0^t u_1(x)R_{S1}(t-x)\,dx, \tag{230}$$

with

- $1-F(t) = \Pr\{\text{duration of the first operating time} > t \mid Z_1 \text{ is entered at } t=0\}$
- $\int_0^t u_1(x)R_{S1}(t-x)\,dx = \Pr\{(S_{RP1} \leq t \cap \text{at the time } t=S_{RP1} \text{ the reserve element is up} \cap \text{system is up in } (S_{RP1},t)) \mid Z_1 \text{ is entered at } t=0\}.$

Considering Fig.26a, the computation of $u_1(x)$ leads to

$$\begin{aligned} u_1(x) &= \lim_{\delta x\to 0} \frac{1}{\delta x} \Pr\{(x<S_{RP1} \leq x+\delta x \cap \text{reserve element is up at } t=x) \mid Z_1 \text{ is entered at } t=0\}, \quad \delta x > 0 \\ &= f(x)PA_d(x), \end{aligned} \tag{231}$$

where $PA_d(x) = \Pr\{\text{reserve element up at } t=x \mid Z_1 \text{ is entered at } t=0\}$ is given by (eq.(145) with $p=0$)

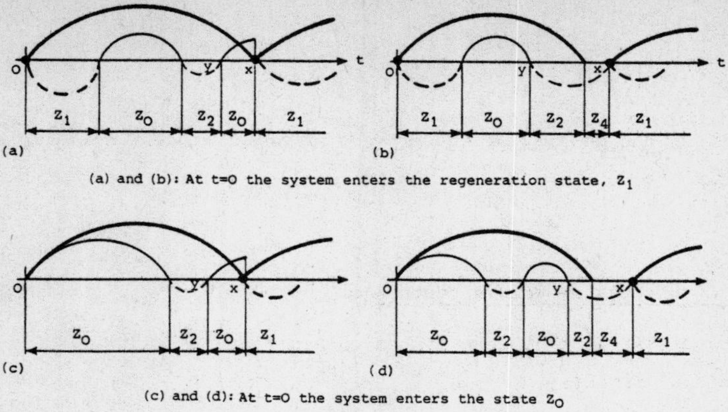

(a) and (b): At t=0 the system enters the regeneration state, Z_1
(c) and (d): At t=0 the system enters the state Z_0

Fig. 26. **Time schedule at the time origin of the model described by Fig.24**

$$PA_d(x) = \int_0^x h'_{dud}(y) e^{-\lambda_r (x-y)} dy, \tag{232}$$

with

$$h'_{dud}(y) = g(y)+g(y)*v(y)*w(y)+g(y)*v(y)*w(y)*v(y)*w(y)+\ldots \; . \tag{233}$$

The point-availability

$$PA_{S1}(t) = \Pr\{\text{system up at } t \mid Z_1 \text{ is entered at } t=0\}, \tag{234}$$

is obtained from

$$PA_{S1}(t) = 1-F(t)+\int_0^t u_1(x) PA_{S1}(t-x)dx+\int_0^t u_2(x) PA_{S1}(t-x)dx, \tag{235}$$

with

- $1-F(t) = \Pr\{\text{duration of the first operating time} > t \mid Z_1 \text{ is entered at } t=0\}$
- $\int_0^t u_1(x) PA_{S1}(t-x)dx = \Pr\{(S_{RP1} \leq t \cap$ at the time $t = S_{RP1}$ the reserve element is up $\cap$ system is up at t$) \mid Z_1$ is entered at $t=0\}$
- $\int_0^t u_2(x) PA_{S1}(t-x)dx = \Pr\{(S_{RP1} \leq t \cap$ system has failed in $(0,S_{RP1}) \cap$ system is up at t$) \mid Z_1$ is entered at $t=0\}$.

The computation of $u_2(x)$ is similar to that of $u_1(x)$ and leads to (Fig.26b)

$$u_2(x) = \lim_{\delta x \to 0} \frac{1}{\delta x} \Pr\{(x < S_{RP1} \leq x+\delta x \cap \text{system has failed in } (0,x)) \mid Z_1 \text{ is entered at } t=0\}, \qquad \delta x > 0$$

$$= g(x)F(x)+\int_0^x h'_{udd}(y) w(x-y) [F(x)-F(y)]dy, \tag{236}$$

with

$$h'_{udd}(y) = g(y)*v(y)+g(y)*v(y)*w(y)*v(y)+\dots \quad (237)$$

$u_1(x)+u_2(x)$ is the density of the *embedded renewal process* defining successive occurrences of state Z_1.

5.2.4.2 At $t=0$ the system enters state Z_0

Let us now assume that at $t=0$, the system enters state Z_0.

The reliability function

$$R_{S0}(t) = \Pr\{\text{system up in } (0,t) \mid Z_0 \text{ is entered at } t=0\} \quad (238)$$

is given by

$$R_{S0}(t) = 1-F(t)+\int_0^t u_3(x)R_{S1}(t-x)dx, \quad (239)$$

with (eq.(230) and Fig.26c))

$$u_3(x) = \lim_{\delta x\to 0} \frac{1}{\delta x}\Pr\{(x<S_{RP1}\leq x+\delta x \cap \text{reserve element is up at } t=x \mid Z_0 \text{ is entered at } t=0\}, \quad \delta x>0$$
$$= f(x)PA_u(x), \quad (240)$$

where $PA_u(x) = \Pr\{\text{reserve element up at } t=x \mid Z_0 \text{ is entered at } t=0\}$ is given by (eq.(145) with $p=1$)

$$PA_u(x) = e^{-\lambda_r x}+\int_0^x h'_{duu}(y)e^{-\lambda_r(x-y)}dy, \quad (241)$$

with

$$h'_{duu}(y) = v(y)*w(y)+v(y)*w(y)*v(y)*w(y)+\dots \quad (242)$$

The point-availability

$$PA_{S0}(t) = \Pr\{\text{system up at } t \mid Z_0 \text{ is entered at } t=0\} \quad (243)$$

is obtained from

$$PA_{S0}(t) = 1-F(t)+\int_0^t u_3(x)PA_{S1}(t-x)dx+\int_0^t u_4(x)PA_{S1}(t-x)dx, \quad (244)$$

with (eq.(235) and Fig.26d))

$$u_4(x) = \lim_{\delta x \to 0} \frac{1}{\delta x} \Pr\{(x < S_{RP1} \leq x+\delta x \cap \text{system has failed in } (0,x)) \mid Z_0 \text{ is entered at } t=0\}, \quad \delta x > 0$$
$$= \int_0^x h'_{udu}(y)w(x-y)[F(x)-F(y)]dy, \tag{245}$$

with

$$h'_{udu}(y) = v(y)+v(y)*w(y)*v(y)+v(y)*w(y)*v(y)*w(y)*v(y)+\ldots \;. \tag{246}$$

5.2.4.3 <u>Solution for some particular cases</u>

The Laplace transforms of equations (230),(235),(239) and (244) are given by

$$\tilde{R}_{S1}(s) = \frac{1-\tilde{f}(s)}{s(1-\tilde{u}_1(s))} \tag{247}$$

$$\tilde{PA}_{S1}(s) = \frac{1-\tilde{f}(s)}{s[1-(\tilde{u}_1(s)+\tilde{u}_2(s))]} \tag{248}$$

$$\tilde{R}_{SO}(s) = \frac{1-\tilde{f}(s)}{s} + \tilde{u}_3(s)\tilde{R}_{S1}(s) \tag{249}$$

$$\tilde{PA}_{SO}(s) = \frac{1-\tilde{f}(s)}{s} + (\tilde{u}_3(s)+\tilde{u}_4(s))\tilde{PA}_{S1}(s). \tag{250}$$

However, difficulties generally arise in the computation of $\tilde{f}(s)$, $\tilde{u}_1(s)$, $\tilde{u}_2(s)$, $\tilde{u}_3(s)$, and $\tilde{u}_4(s)$, as well as in the inversion of the final Laplace transforms. A closed expression can be found for the mean time-to-system-failure and for the stationary-state point-availability and average-availability. From equations (247) through (249) it follows that

$$MTTF_{S1} = \tilde{R}_{S1}(0) = \frac{MTTF}{1-\int_0^\infty u_1(x)dx} \tag{251}$$

$$MTTF_{SO} = \tilde{R}_{SO}(0) = MTTF\left[1+\frac{\int_0^\infty u_3(x)dx}{1-\int_0^\infty u_1(x)dx}\right] \tag{252}$$

$$\lim_{t\to\infty} PA_{S1}(t) = \lim_{s\to 0} s\,\tilde{PA}_{S1}(s) = PA_S = AA_S = \frac{MTTF}{\int_0^\infty x(u_1(x)+u_2(x))dx}, \tag{253}$$

with

$$MTTF = \int_0^\infty (1-F(x))dx. \tag{254}$$

The model of Fig.24 contains the models of Fig.13, Fig.22, and the 1-out-of-2 standby redundancy, with arbitrary failure and repair rates as particular cases. For the standby redundancy, with failure-free times distributed according to $F(x)$ and repair times distributed according to $G(x)$, one obtains $u_1(x) = f(x)G(x)$, $u_2(x) = g(x)\cdot F(x)$, $u_3(x) = f(x)$ and $u_4(x) \equiv 0$, and thus

$$\tilde{R}_{SO}(s) = \frac{1-\tilde{f}(s)}{s} + \frac{\tilde{f}(s)(1-\tilde{f}(s))}{s[1-\tilde{u}_1(s)]} \tag{255}$$

$$\tilde{PA}_{SO}(s) = \frac{1-\tilde{f}(s)}{s} + \frac{\tilde{f}(s)(1-\tilde{f}(s)}{s[1-\tilde{u}_1(s)-\tilde{u}_2(s)]}, \tag{256}$$

with

$$\tilde{u}_1(s) = \int_0^\infty f(t)G(t)e^{-st}dt \tag{257}$$

and

$$\tilde{u}_2(s) = \int_0^\infty g(t)F(t)e^{-st}dt. \tag{258}$$

The mean time-to-system-failure and the stationary-state point-availability and average-availability are then given by (eq.(251) to (253))

$$MTTF_{S1} = \frac{MTTF}{1-\int_0^\infty f(x)G(x)dx} \tag{259}$$

$$MTTF_{SO} = MTTF[1+ \frac{1}{1-\int_0^\infty f(x)G(x)dx}] \tag{260}$$

$$PA_S = AA_S = \frac{MTTF}{\int_0^\infty xd(F(x)G(x))}, \tag{261}$$

with MTTF as in equation (254). For $F(x) = 1-e^{-\lambda x}$, equations (260) and (261) lead to equations (219) and (223) with $\lambda_r = 0$.

The main results obtained for the 1-out-of-2 redundancies are summarized in Table 10.

5.3 k-out-of-n redundancies

By a k-out-of-n redundancy, one generally means a reliability structure consisting of n identical elements, k of which are necessary for the required function and n-k which stay in the reserve state. Assuming ideal failure detection and switching, the reliability block diagram can be represented as in Fig.27 (p.70).

			1-out-of-2 standby redundancy	1-out-of-2 warm redundancy		1-out-of-2 active redundancy
Elements E_1, E_2	Distribution of the failure-free times	in the operating state	$F(t)$	$1-e^{-\lambda t}$	$F(t)$	$1-e^{-\lambda t}$
		in the reserve state	——	$1-e^{-\lambda_r t}$	$1-e^{-\lambda_r t}$	$1-e^{-\lambda t}$
	Distribution of the repair times	in the operating state	$G(t)$	$G(t)$	$G(t)$	$G(t)$
		in the reserve state	——	$G(t)$	$W(t)$	$G(t)$
	Mean time	to-failure	$MTTF=\int_0^\infty (1-F(t))dt$	$1/\lambda$ bzw. $1/\lambda_r$	MTTF bzw. $1/\lambda_r$	$1/\lambda$
		to-repair	$MTTR=\int_0^\infty (1-G(t))dt$	MTTR	MTTR bzw. $MTTR_W$	MTTR
1-out-of-2 redundancy	Mean time-to-system-failure for the case in which the system is new at $t=0$ ($MTTF_{SO}$)		$MTTF + \frac{MTTF}{1-\int_0^\infty f(t)G(t)dt}$	$\frac{1}{\lambda}+\frac{1}{(\lambda+\lambda_r)(1-\tilde{g}(\lambda))} \approx \frac{1}{\lambda}(1+\frac{1}{(\lambda+\lambda_r)MTTR})$ ◆	$MTTF + \frac{MTTF\int_0^\infty u_3(t)dt}{1-\int_0^\infty u_1(t)dt}$	$\frac{1}{\lambda}+\frac{1}{2\lambda(1-\tilde{g}(\lambda))} \approx \frac{1}{\lambda}(1+\frac{1}{2\lambda MTTR})$ ◆
	Stationary-state point-availability and average-availability ($PA_S = AA_S$)		$\frac{MTTF}{\int_0^\infty t\,d(F(t)G(t))}$	$\frac{\lambda+\lambda_r(1-\tilde{g}(\lambda))}{\lambda(\lambda+\lambda_r)MTTR+\lambda\tilde{g}(\lambda)}$	$\frac{MTTF}{\int_0^\infty t(u_1(t)+u_2(t))dt}$	$\frac{2-\tilde{g}(\lambda)}{2\lambda MTTR+\tilde{g}(\lambda)}$
	Stationary-state interval-reliability ($IR_S(\theta)$)		$\approx R_{SO}(\theta)$ ◇	$\approx R_{SO}(\theta)$ ◇	$\approx R_{SO}(\theta)$ ◇	$\approx R_{SO}(\theta)$ ◇

◇ $R_{SO}(t) = Pr\{\text{system up in } (0,t) \mid \text{state } Z_0 \text{ is entered at } t=0\}$; ◆ for $\lambda\,MTTR \ll 1$ follows $\tilde{g}(\lambda) \approx 1-\lambda\,MTTR$

Table 10. **Main results for repairable 1-out-of-2 redundancies (only one repair crew)**

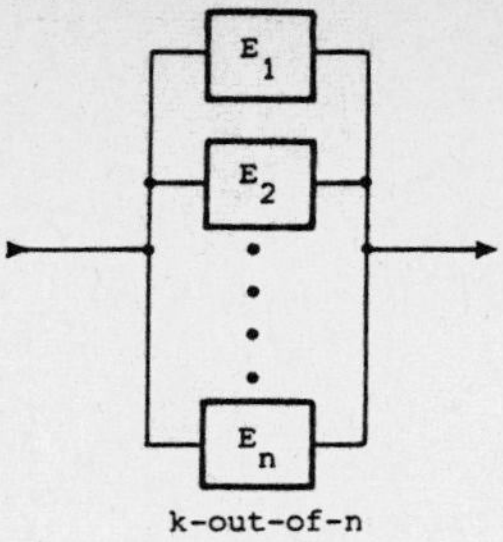

Fig. 27. Reliability block diagram for a k-out-of-n redundancy

5.3.1 Constant failure and repair rates

For the investigation let us first assume that the system consists of n identical elements having constant failure and repair rates, i.e. with distribution functions as in equations (204) to (206). In this case, the stochastic behaviour is described by a time-homogeneous Markov process, more precisely by a *birth and death process*. Assuming as in sections 5.1 and 5.2 that the system has only one repair crew and that no further failure can occur at system down one obtains the diagram of Fig.28 for a transition in an arbitrary time interval $(t,t+\delta t)$. In state Z_i there are i elements

$\rho_{i(i+1)} = \nu_i$, $i=0,1,\ldots,n-k$; $\rho_{i(i-1)} = \mu$, $i=1,2,\ldots,n-k+1$
Failure rates λ, λ_r; repair rate μ; $\nu_i = k\lambda + (n-k-i)\lambda_r$

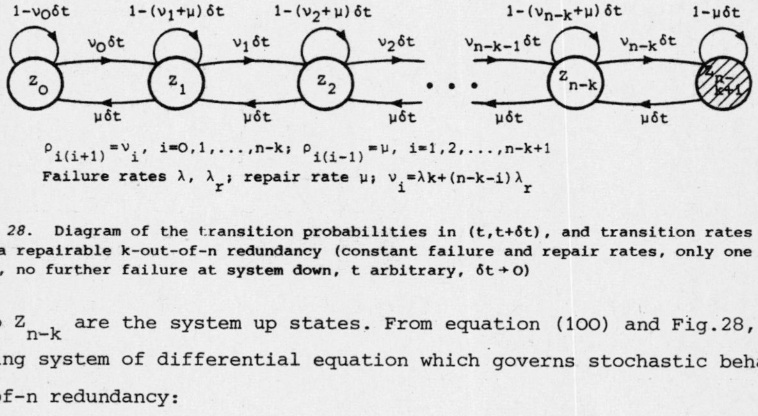

Fig. 28. Diagram of the transition probabilities in $(t,t+\delta t)$, and transition rates (ρ_{ij}), for a repairable k-out-of-n redundancy (constant failure and repair rates, only one repair crew, no further failure at system down, t arbitrary, $\delta t \to 0$)

down. Z_0 to Z_{n-k} are the system up states. From equation (100) and Fig.28, one obtains the following system of differential equation which governs stochastic behaviour of the k-out-of-n redundancy:

$$\dot{P}_0(t) = -\nu_0 P_0(t) + \mu P_1(t)$$
$$\dot{P}_i(t) = \nu_{i-1} P_{i-1}(t) - (\nu_i + \mu) P_i(t) + \mu P_{i+1}(t), \quad i = 1,2,\ldots,n-k$$
$$\dot{P}_{n-k+1}(t) = \nu_{n-k} P_{n-k}(t) - \mu P_{n-k+1}(t),$$

(262)

with

$$\nu_i = k\lambda + (n-k-i)\lambda_r, \quad i = 0,1,2,\ldots,n-k.$$

The reliability functions $R_{Si}(t)$ (eq.(106)) are obtained from equation (262) by considering equations (107) to (110), or directly from equation (111). The solution according to equation (111) leads to the following system of integral equations

$$R_{SO}(t) = e^{-\nu_O t}+\int_O^t \nu_O e^{-\nu_O x} R_{S1}(t-x)dx$$

$$R_{Si}(t) = e^{-(\nu_i+\mu)t}+\int_O^t (\nu_i R_{Si+1}(t-x)+\mu R_{Si-1}(t-x))e^{-(\nu_i+\mu)x}dx, \quad i=1,2,\ldots,n-k-1$$

$$R_{Sn-k}(t) = e^{-(\nu_{n-k}+\mu)t}+\int_O^t \mu R_{Sn-k-1}(t-x)e^{-(\nu_{n-k}+\mu)x}dx, \tag{264}$$

which can be solved using Laplace transforms (eq.(28)). The functions $R_{Si}(t)$ are dependent upon the quantity n-k. Assuming that the system enters state Z_O at $t=O$, one obtains, for $n-k=1$,

$$\tilde{R}_{SO_1}(s) = \frac{s+\nu_O+\nu_1+\mu}{(s+\nu_O)(s+\nu_1)+s\mu} \tag{265}$$

and, for n-k=2,

$$\tilde{R}_{SO_2}(s) = \frac{(s+\nu_O+\nu_1+\mu)(s+\nu_2+\mu)+\nu_1(\nu_O-\mu)}{s(s+\nu_O+\nu_1+\mu)(s+\nu_2+\mu)+\nu_O\nu_1\nu_2+s\nu_1(\nu_O-\mu)} . \tag{266}$$

The mean times-to-system-failure $MTTF_{Si}$ are the solution of the following system of algebraic equations (eq.(119))

$$MTTF_{SO} = \frac{1}{\nu_O} + MTTF_{S1}$$

$$MTTF_{Si} = \frac{1}{\nu_i+\mu}(1+\nu_i MTTF_{Si+1}+\mu MTTF_{Si-1}), \quad i=1,2,\ldots,n-k-1$$

$$MTTF_{Sn-k} = \frac{1}{\nu_{n-k}+\mu}(1+\mu MTTF_{Sn-k-1}), \tag{267}$$

with ν_i as in equation (263). Assuming that the system enters state Z_O at $t=O$ it follows, for n-k=1,

$$MTTF_{SO_1} = \frac{\nu_O+\nu_1+\mu}{\nu_O\nu_1} , \tag{268}$$

and, for n-k=2,

$$MTTF_{SO_2} = \frac{\nu_2(\nu_O+\nu_1+\mu)+\mu(\nu_O+\mu)+\nu_O\nu_1}{\nu_O\nu_1\nu_2} . \tag{269}$$

The point-availabilities $PA_{Si}(t)$ (eq.(104)) can be computed from equation (262) by considering equations (100) to (105), or from equations (122) and (105) with $Q_{ij}(x)$ as in equation (90) and ρ_{ij} as in Fig.28.

The interval reliabilities

$$IR_{Si}(t,t+\theta) = Pr\{\text{system up in } (t,t+\theta) \mid Z_i \text{ is entered at } t=0\}, \quad Z_i \in M \tag{270}$$

are given by

$$IR_{Si}(t,t+\theta) = \sum_{Z_j \in M} P_{ij}(t) R_{Sj}(\theta), \tag{271}$$

where $P_{ij}(t)$ and $R_{Sj}(\theta)$ are obtained from equations (122) and (264), respectively. M is the set of up states, i.e. $M = \{Z_0, Z_1, \ldots, Z_{n-k}\}$.

Asymptotic behaviour exists, independend of initial conditions at $t=0$, and leads to

$$\lim_{t \to \infty} PA_{Si}(t) = PA_S = \sum_{i=0}^{n-k} p_i = 1 - p_{n-k+1}, \tag{272}$$

where p_i are the limits for $t \to \infty$ of the state probabilities $P_i(t)$ given by equation (262). $p_0, p_1, \ldots, p_{n-k+1}$ are obtained by solving the following system of algebraic equations (eq.(262) with $\dot{P}_i(t) = 0$ and $P_i(t) = p_i$)

$$\begin{aligned} 0 &= -\nu_0 p_0 + \mu p_1 \\ 0 &= \nu_{i-1} p_{i-1} - (\nu_i + \mu) p_i + \mu p_{i+1}, \qquad i = 1,2,\ldots,n-k \\ 0 &= \nu_{n-k} p_{n-k} - \mu p_{n-k+1} \\ & p_0 + p_1 + p_2 + \ldots + p_{n-k+1} = 1, \end{aligned} \tag{273}$$

and given by

$$p_i = \frac{\pi_i}{\sum_{j=0}^{n-k+1} \pi_j} \qquad \text{with } \pi_0 = 1 \text{ and } \pi_j = \frac{\nu_0 \nu_1 \nu_2 \ldots \nu_{j-1}}{\mu^j} . \tag{274}$$

Expression (272) also gives the stationary-state point-availability and average-availability (eq.(156)). This remark holds also for equation (271), with p_j instead of $P_{ij}(t)$.

With $\lambda_r \equiv 0$ and $\lambda_r = \lambda$ one obtains the results for, respectively, the *standby* and the *active redundancy* cases.

Table 11 summarizes the main results obtained for k-out-of-n redundancies.

5.3.2 Constant failure rates and arbitrary repair rate

Generalization of the repair rate leads to a *regenerative stochastic process with only two regeneration states*, states Z_0 and Z_1 in Fig.28. The investigation is similar

		Mean time-to-system-failure for the case in which the system enters state Z_O at t=O ($MTTF_{SO}$)	Stationary-state point-availability and average-availability ($PA_S = AA_S$)	Stationary-state interval-reliability ($IR_S(\theta)$)
n - k = 1	general case	$\frac{\nu_O+\nu_1+\mu}{\nu_O\nu_1}$	$\frac{\nu_O\mu+\mu^2}{\nu_O\nu_1+\nu_O\mu+\mu^2}$	$\frac{\mu^2 R_{SO}(\theta)+\nu_O\mu R_{S1}(\theta)}{\nu_O\nu_1+\nu_O\mu+\mu^2} \approx R_{SO}(\theta)$
	n = 2 k = 1	$\frac{2\lambda+\lambda_r+\mu}{\lambda(\lambda+\lambda_r)}$	$\frac{\mu(\lambda+\lambda_r+\mu)}{(\lambda+\lambda_r)(\lambda+\mu)+\mu^2}$	$\frac{\mu^2 R_{SO}(\theta)+\mu(\lambda+\lambda_r)R_{S1}(\theta)}{(\lambda+\lambda_r)(\lambda+\mu)+\mu^2} \approx R_{SO}(\theta)$
	n = 3 k = 2	$\frac{4\lambda+\lambda_r+\mu}{2\lambda(2\lambda+\lambda_r)}$	$\frac{\mu(2\lambda+\lambda_r+\mu)}{(2\lambda+\lambda_r)(2\lambda+\mu)+\mu^2}$	$\frac{\mu^2 R_{SO}(\theta)+\mu(2\lambda+\lambda_r)R_{S1}(\theta)}{(2\lambda+\lambda_r)(2\lambda+\mu)+\mu^2} \approx R_{SO}(\theta)$
n - k = 2	general case	$\frac{\nu_2(\nu_O+\nu_1+\mu)+\mu(\nu_O+\mu)+\nu_O\nu_1}{\nu_O\nu_1\nu_2}$	$\frac{\nu_O\nu_1\mu+\nu_O\mu^2+\mu^3}{\nu_O\nu_1\nu_2+\nu_O\nu_1\mu+\nu_O\mu^2+\mu^3}$	$\frac{\mu^3 R_{SO}(\theta)+\mu^2\nu_O R_{S1}(\theta)+\mu\nu_O\nu_1 R_{S2}(\theta)}{\nu_O\nu_1\nu_2+\nu_O\nu_1\mu+\nu_O\mu^2+\mu^3} \approx R_{SO}(\theta)$
	n = 3 k = 1	$\frac{1}{\lambda}+\frac{\lambda(2\lambda+3\lambda_r+\mu)+\mu(\lambda+2\lambda_r+\mu)}{\lambda(\lambda+\lambda_r)(\lambda+2\lambda_r)}$	$\frac{\mu[(\lambda+2\lambda_r)(\lambda+\lambda_r)+\mu(\lambda+2\lambda_r)+\mu^2]}{(\lambda+2\lambda_r)[\lambda(\lambda+\lambda_r)+\mu(\lambda+\lambda_r)+\mu^2]+\mu^3}$	$\frac{\mu^3 R_{SO}(\theta)+\mu^2(\lambda+2\lambda_r)R_{S1}(\theta)+\mu(\lambda+2\lambda_r)(\lambda+\lambda_r)R_{S2}(\theta)}{(\lambda+2\lambda_r)[\lambda(\lambda+\lambda_r)+\mu(\lambda+\lambda_r)+\mu^2]+\mu^3} \approx R_{SO}(\theta)$
	n = 5 k = 3	$\frac{1}{3\lambda}+\frac{3\lambda(6\lambda+3\lambda_r+\mu)+\mu(3\lambda+2\lambda_r+\mu)}{(3\lambda+2\lambda_r)(3\lambda+\lambda_r)3\lambda}$	$\frac{\mu(3\lambda+2\lambda_r)(3\lambda+\lambda_r+\mu)+\mu^3}{(3\lambda+2\lambda_r)(3\lambda+\lambda_r)(3\lambda+\mu)+\mu^2(3\lambda+2\lambda_r)+\mu^3}$	$\frac{\mu^3 R_{SO}(\theta)+\mu^2(3\lambda+2\lambda_r)R_{S1}(\theta)+\mu(3\lambda+2\lambda_r)(3\lambda+\lambda_r)R_{S2}(\theta)}{(3\lambda+2\lambda_r)(3\lambda+\lambda_r)(3\lambda+\mu)+\mu^2(3\lambda+2\lambda_r)+\mu^3} \approx R_{SO}(\theta)$
n - k arbitrary		$MTBF_{Sj} = \int_O^\infty R_{Sj}(t)\,dt$	$PA_S = AA_S = \sum_{j=O}^{n-k} p_j$	$IR_S(\theta) = \sum_{j=O}^{n-k} p_j R_{Sj}(\theta)$
			$p_j = \pi_j / \sum_{i=O}^{n-k+1} \pi_i$ with $\pi_O = 1$ and $\pi_i = \frac{\nu_O\nu_1\ldots\nu_{i-1}}{\mu^i}$	

$R_{Sj}(t)$ according to equation (264); $\nu_i = k\lambda+(n-k-i)\lambda_r$, $i = O,1,\ldots,n-k$; λ,λ_r = failure rates ($\lambda_r=\lambda \rightarrow$ active, $\lambda_r \equiv O \rightarrow$ stanby); μ = repair rate

Table 11. Main results for repairable k-out-of-n redundancies (constant failure and repair rates, one repair crew, no further failure at system down)

to that of the model of Fig.22. As an example, let us consider a *2-out-of-3 active redundancy* with failure-free times distributed according to equation (204) and repair times according to $G(x)$, with density $g(x)$. The investigation is performed by assuming that the system enters state Z_0 at $t=0$, and by making use of the regenerative property of states Z_0 and Z_1.

For the reliability function $R_{S0}(t)$ (eq.(181)) one obtains

$$R_{S0}(t) = e^{-3\lambda t}+\int_0^t 3\lambda e^{-3\lambda x} e^{-2\lambda(t-x)}(1-G(t-x))dx$$
$$+\int_0^t\int_0^y 3\lambda e^{-3\lambda x} g(y-x)e^{-2\lambda(y-x)}R_{S0}(t-y)dxdy. \tag{275}$$

$R_{S0}(t)$ is the probability of the three events shown in Fig.29a. The Laplace transform (eq.(28)) of $R_{S0}(t)$ is given by

$$\tilde{R}_{S0}(s) = \frac{s+5\lambda-3\lambda\tilde{g}(s+2\lambda)}{(s+2\lambda)(s+3\lambda)-3\lambda(s+2\lambda)\tilde{g}(s+2\lambda)}, \tag{276}$$

and the mean time-to-system-failure $MTTF_{S0}$ by

$$MTTF_{S0} = \frac{5-3\tilde{g}(2\lambda)}{6\lambda(1-\tilde{g}(2\lambda))}. \tag{277}$$

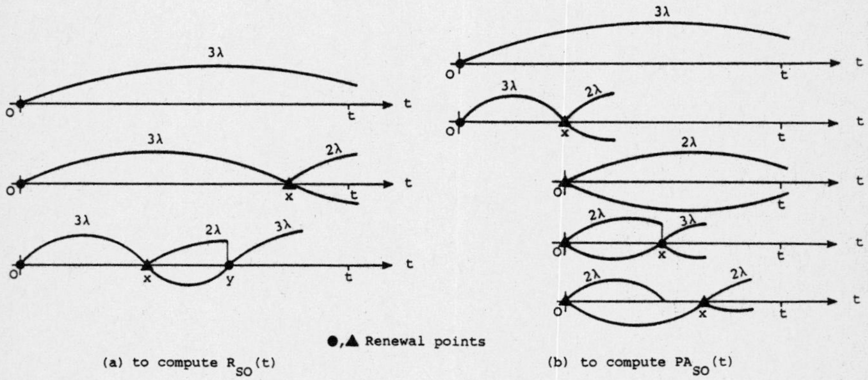

Fig. 29. **Time schedule of a repairable 2-out-of-3 active redundancy (constant failure rates, arbitrary repair rate, only one repair crew, no further failure at system down)**

For computation of the point-availability $PA_{S0}(t)$ (eq.(184)), one must consider the events shown in Fig.29b. This leads to the following system of integral equations:

$$PA_{SO}(t) = e^{-3\lambda t} + \int_0^t 3\lambda e^{-3\lambda x} PA_{S1}(t-x)\,dx$$

$$PA_{S1}(t) = e^{-2\lambda t}(1-G(t)) + \int_0^t g(x) e^{-2\lambda x} PA_{SO}(t-x)\,dx$$

$$+ \int_0^t g(x)(1-e^{-2\lambda x}) PA_{S1}(t-x)\,dx. \tag{278}$$

The Laplace transform (eq.(28)) of $PA_{SO}(t)$ is then given by

$$\tilde{PA}_{SO}(s) = \frac{(s+2\lambda)[1+\tilde{g}(s+2\lambda)-\tilde{g}(s)]+3\lambda(1-\tilde{g}(s+2\lambda))}{s(s+2\lambda)[1+\tilde{g}(s+2\lambda)-\tilde{g}(s)]+3\lambda(s+2\lambda)(1-\tilde{g}(s))}. \tag{279}$$

The interval-reliability $IR_{SO}(t,t+\theta)$ (eq.(186)) can be approximated using an expression similar to that of equation (213). The approximation is good if $MTTR \ll 1/\lambda$ holds, with MTTR as in equation (222).

The asymptotic behaviour exists, independent of the initial conditions at $t=0$, and leads to

$$\lim_{t\to\infty} PA_{SO}(t) = PA_S = \frac{3-\tilde{g}(2\lambda)}{2\tilde{g}(2\lambda)+6\lambda MTTR}, \tag{280}$$

with MTTR as in equation (222). Expression (280) also gives stationary-state point-availability and average-availability (eq.(156)).

As in the case of the 1-out-of-2 redundancy of section 5.2, generalization of repair and failure rates leads to a *non-regenerative stochastic process.*

5.4 Series/parallel structures

A combination of series and parallel structures leads to series/parallel structures (Table 2 on p.6). As an example, let us consider a majority redundancy with $n=1$, i.e. a *2-out-of-3 active redundancy in series with a voter.*

5.4.1 Constant failure and repair rates

For the investigation, let us first consider the case for which the distribution functions are:

$$F(x) = 1-e^{-\lambda x} \tag{204}$$

for the failure-free times of each of the three elements in redundancy,

$$F_V(x) = 1-e^{-\lambda_V x} \tag{281}$$

for the failure-free times of the voter, and

$$G(x) = 1-e^{-\mu x} \tag{206}$$

for the repair times. As in sections 5.1 to 5.3, the system has only one repair crew and no further failure can occur at system down. The stochastic behaviour is described by a *time-homogeneous Markov process* with 5 states. Fig.30 gives the reliability block diagram and the corresponding transition probabilities diagram for an arbitrary time interval $(t,t+\delta t)$. States Z_0 and Z_1 are the system up states.

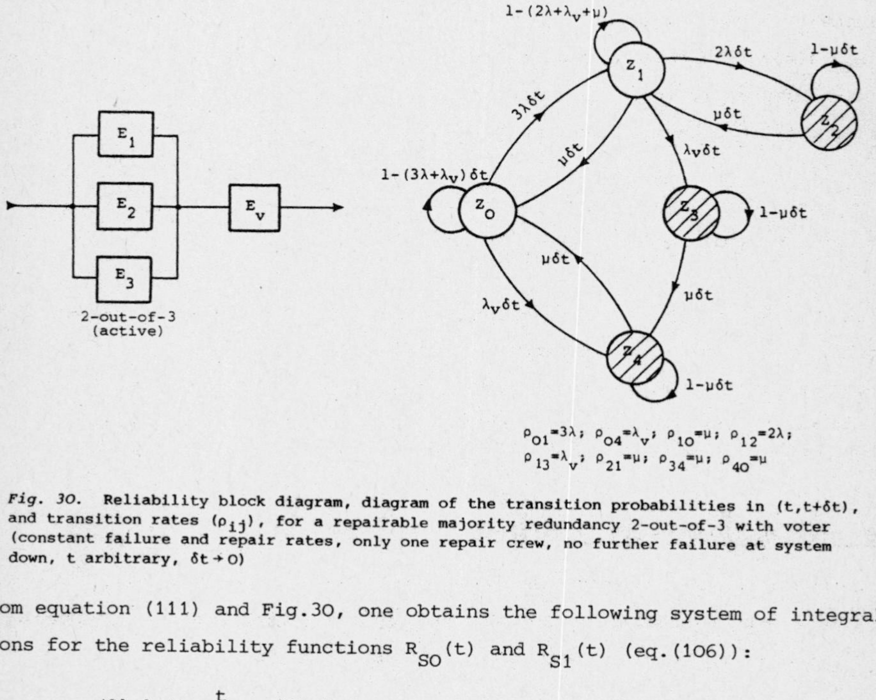

Fig. 30. **Reliability block diagram, diagram of the transition probabilities in $(t,t+\delta t)$, and transition rates (ρ_{ij}), for a repairable majority redundancy 2-out-of-3 with voter (constant failure and repair rates, only one repair crew, no further failure at system down, t arbitrary, $\delta t \to 0$)**

From equation (111) and Fig.30, one obtains the following system of integral equations for the reliability functions $R_{S0}(t)$ and $R_{S1}(t)$ (eq.(106)):

$$R_{S0}(t) = e^{-(3\lambda+\lambda_v)t}+\int_0^t 3\lambda e^{-(3\lambda+\lambda_v)x} R_{S1}(t-x)\,dx$$

$$R_{S1}(t) = e^{-(2\lambda+\lambda_v+\mu)t}+\int_0^t \mu e^{-(2\lambda+\lambda_v+\mu)x} R_{S0}(t-x)\,dx. \tag{282}$$

The Laplace transform of $R_{S0}(t)$ and the mean time-to-system-failure $MTTF_{S0}$ are then given by

$$\tilde{R}_{S0}(s) = \frac{s+5\lambda+\lambda_v+\mu}{(s+3\lambda+\lambda_v)(s+2\lambda+\lambda_v)+\mu(s+\lambda_v)} \tag{283}$$

and

$$MTTF_{S0} = \frac{5\lambda+\lambda_v+\mu}{(3\lambda+\lambda_v)(2\lambda+\lambda_v)+\mu\lambda_v}\,. \tag{284}$$

The point-availabilities $PA_{Si}(t)$ (eq.(104)) can be computed from equations (100) to (105) with ρ_{ij} as in Fig.30, or from equations (122) and (105) with $Q_{ij}(x)$ as in equation (90) and ρ_{ij} as in Fig.30.

The interval reliabilities $IR_{Si}(t,t+\theta)$ (eq.(270)) are given by equation (271) with $M=\{Z_0,Z_1\}$, $P_{ij}(t)$ from equation (122) and $R_{Sj}(\theta)$ from equation (282).

Asymptotic behaviour exists, independent of the initial conditions at $t=0$, and leads to

$$\lim_{t\to\infty} PA_{S0}(t) = PA_S = \frac{\mu(3\lambda+\lambda_v+\mu)}{(3\lambda+\lambda_v+\mu)(\mu+\lambda_v)+3\lambda(2\lambda+\lambda_v)} \tag{285}$$

$$\lim_{t\to\infty} IR_{S0}(t,t+\theta) = IR_S(\theta) = \frac{(\lambda_v+\mu)\mu R_{S0}(\theta)+3\lambda\mu R_{S1}(\theta)}{(3\lambda+\lambda_v+\mu)(\mu+\lambda_v)+3\lambda(2\lambda+\lambda_v)}\ , \tag{286}$$

with $R_{S0}(\theta)$ and $R_{S1}(\theta)$ from equation (282). Expression (285) also gives stationary-state point-availability and average-availability (eq.(156)). This remark holds also for equation (286).

5.4.2 Constant failure rates and arbitrary repair rate

Generalization of the repair times distribution function, say according to $G(x)$, with density $g(x)$, leads to a *regenerative process with only three regeneration states*, states Z_0, Z_1 and Z_4 in Fig.30. These three states constitute an *embedded semi-Markov process* on which the analysis can be based. The corresponding semi-Markov transition probabilities $Q_{ij}(x)$ (eq.(87)) are given by

$$Q_{01}(x) = \frac{3\lambda}{3\lambda+\lambda_v}\,(1-e^{-(3\lambda+\lambda_v)x})$$

$$Q_{10}(x) = \int_0^x g(y)e^{-(2\lambda+\lambda_v)y}dy$$

$$Q_{12}(x) = \int_0^x 2\lambda e^{-(2\lambda+\lambda_v)y}(1-G(y))dy$$

$$Q_{121}(x) = \int_0^x \frac{2\lambda}{2\lambda+\lambda_v}\,(1-e^{-(2\lambda+\lambda_v)y})g(y)dy$$

$$Q_{13}(x) = \frac{\lambda_v}{2\lambda}\,Q_{12}(x)$$

$$Q_{134}(x) = \frac{\lambda_v}{2\lambda}\,Q_{121}(x)$$

$$Q_{04}(x) = \frac{\lambda_v}{3\lambda}\,Q_{01}(x)$$

$$Q_{40}(x) = G(x). \tag{287}$$

$Q_{12}(x)$ and $Q_{13}(x)$ will be used to compute the reliability function. $Q_{121}(x)$ and $Q_{134}(x)$ take care of a transition throughout the *non-regenerative* states Z_2 and Z_3, respectively. They will be used to compute the point-availability.

For the reliability functions $R_{SO}(t)$ and $R_{S1}(t)$ one obtains from equations (111) and (287)

$$R_{SO}(t) = e^{-(3\lambda+\lambda_v)t} + \int_0^t 3\lambda e^{-(3\lambda+\lambda_v)x} R_{S1}(t-x)\,dx$$

$$R_{S1}(t) = e^{-(2\lambda+\lambda_v)t}(1-G(t)) + \int_0^t g(x) e^{-(2\lambda+\lambda_v)x} R_{SO}(t-x)\,dx. \qquad (288)$$

The Laplace transform (eq. (28)) of $R_{SO}(t)$ is then given by

$$\tilde{R}_{SO}(s) = \frac{s+5\lambda+\lambda_v-3\lambda\tilde{g}(s+2\lambda+\lambda_v)}{(s+2\lambda+\lambda_v)[s+\lambda_v+3\lambda(1-\tilde{g}(s+2\lambda+\lambda_v))]} \qquad (289)$$

and the mean time-to-system-failure $MTTF_{SO}$ by

$$MTTF_{SO} = \frac{5\lambda+\lambda_v-3\lambda\tilde{g}(2\lambda+\lambda_v)}{(2\lambda+\lambda_v)[\lambda_v+3\lambda(1-\tilde{g}(2\lambda+\lambda_v))]}\,. \qquad (290)$$

The point-availabilities $PA_{Si}(t)$ (eq. (104)) can be computed from equation (105) with $M = \{Z_0, Z_1\}$ and $P_{ij}(t)$ from the following system of integral equations (eq. (122) and (287)):

$$P_{00}(t) = e^{-(3\lambda+\lambda_v)t} + \int_0^t 3\lambda e^{-(3\lambda+\lambda_v)x} P_{10}(t-x)\,dx + \int_0^t \lambda_v e^{-(3\lambda+\lambda_v)x} P_{40}(t-x)\,dx$$

$$P_{10}(t) = \int_0^t g(x) e^{-(2\lambda+\lambda_v)x} P_{00}(t-x)\,dx + \int_0^t \frac{2\lambda}{2\lambda+\lambda_v}(1-e^{-(2\lambda+\lambda_v)x}) g(x)\, P_{10}(t-x)\,dx$$
$$+ \int_0^t \frac{\lambda_v}{2\lambda+\lambda_v}(1-e^{-(2\lambda+\lambda_v)x}) g(x)\, P_{40}(t-x)\,dx$$

$$P_{40}(t) = \int_0^t g(x) P_{00}(t-x).$$

$$P_{01}(t) = \int_0^t 3\, e^{-(3\lambda+\lambda_v)x} P_{11}(t-x)\,dx + \int_0^t \lambda_v e^{-(3\lambda+\lambda_v)x} P_{41}(t-x)\,dx$$

$$P_{11}(t) = e^{-(2\lambda+\lambda_v)t}(1-G(t)) + \int_0^t g(x) e^{-(2\lambda+\lambda_v)x} P_{01}(t-x)\,dx$$
$$+ \int_0^t \frac{1}{2\lambda+\lambda_v}(1-e^{-(2\lambda+\lambda_v)x}) g(x)(2\lambda P_{11}(t-x)+\lambda_v P_{41}(t-x))\,dx$$

$$P_{41}(t) = \int_0^t g(x) P_{01}(t-x)\,dx. \qquad (291)$$

The interval-reliability $IR_{SO}(t,t+\theta)$ can be approximated using equation (213) with $P_{00}(t)$ and $P_{01}(t)$ from equation (291), and $R_{SO}(\theta)$ and $R_{S1}(\theta)$ from equation (288). The approximation, which assumes that state Z_1 is regenerative at each time point, is good if $MTTR \ll 1/\lambda$ holds, with MTTR as in equation (222).

Asymptotic behaviour exists, independent of initial conditions at $t=0$, and leads to

$$\lim_{t\to\infty} PA_{SO}(t) = PA_S = \frac{2\lambda+\lambda_V+\lambda(1-\tilde{g}(2\lambda+\lambda_V))}{(2\lambda+\lambda_V)(1+(3\lambda+\lambda_V)MTTR)+\lambda(\lambda_V MTTR-2)(1-\tilde{g}(2\lambda+\lambda_V))} \tag{292}$$

$$\lim_{t\to\infty} IR_{SO}(t,t+\theta) = IR_S(\theta) \approx \frac{[2\lambda+\lambda_V)-2\lambda(1-\tilde{g}(2\lambda+\lambda_V))]R_{SO}(\theta)}{(2\lambda+\lambda_V)(1+(3\lambda+\lambda_V)MTTR)+\lambda(\lambda_V MTTR-2)(1-\tilde{g}(2\lambda+\lambda_V))}\,. \tag{293}$$

Equation (293) only considers the first term of equation (213). This approximation holds for $MTTR \ll 1/\lambda$.

Generalization of repair and failure rates leads to a *non-regenerative stochastic process*.

CHAPTER 6

APPLICATIONS TO REPAIRABLE SYSTEMS OF COMPLEX STRUCTURE AND TO SPECIAL TOPICS

This chapter considers some important aspects of the investigation of repairable complex structures. Also discussed are the influences of preventive maintenance and imperfect switching on system reliability and availability.

6.1 Repairable systems having complex structure

From a reliability point of view, a structure is complex if its reliability block diagram either does not exist or cannot be reduced to a series/parallel structure. As pointed out in section 2.1, the reliability block diagram does not exist if elements with more than one failure mode (short, open, drift) or more than two states (good/failed) must be considered. In such cases, investigations are generally problem-oriented, and are performed together with an FMECA (Table 5 on p. 16) in order to deal with secondary failures and criticality.

If the reliability block diagram (RBD) exists, but cannot be reduced to a series/parallel structure because of its topology or because of elements which appear twice, investigations are performed by making use of one or more of the following assumptions:

1. Each element of the RBD has a constant failure rate.
2. The flow of failures constitutes a Poisson process.
3. Each element of the RBD has constant failure and repair rates.
4. No further failure can occur at system down.
5. After each repair, the system is good-as-new.
6. Failure-free and repair times are statistically independent.
7. The system has only one repair crew.
8. Each element of the RBD works, fails, and is repaired independently of all other elements.
9. Failure detection is 100% reliable and no hidden failures are present.
10. For each element of the RBD, the mean time-to-repair is much shorter than the mean time-to-failure (MTTR $\ll$ MTTF).
11. Preventive maintenance is not considered.
12. Switching effects are negligible.

One normally assumes that, for each element of the reliability block diagram, only two states (good/failed) and only one failure mode (short or open) are considered, and that the required function is time invariant. With assumptions 3 and 6, the stochastic behaviour of the system is described by a *time-homogeneous Markov process* with a finite state space. Difficulties generally arise with large systems because of the great number of transition probabilities which are involved. Assumption 5 is satisfied if assumption 1 holds. Assumption 2 can be used with large systems [22,28, 36], often even with non-homogeneous Poisson processes [162]. Assumption 4 is true in many practical applications, in particular when assumption 10 holds; it simplifies the availability and interval-reliability analyses (chapter 5). Assumption 7 is true in many practical applications; it simplifies the investigations and is sufficient to assure, together with assumption 1, that the process involved has an *embedded semi-Markov process*. With assumption 8, point-availability can be evaluated using the expression obtained for the reliability function in the non-repairable case. Assumption 10 is generally true; it leads to useful approximations obtained by series expansion (eq.(227) and (228)) or as limit expressions [22-27,51,56-59]. Assumptions 11 and 12 simplify the investigations.

Complex structures of different kinds have been investigated in [1,3,8,9,21,26, 29,34,44,48,49,50,52,62,158-231]. To give an idea of a moderately complex system, let us consider the functional block diagram shown in Fig.31a.

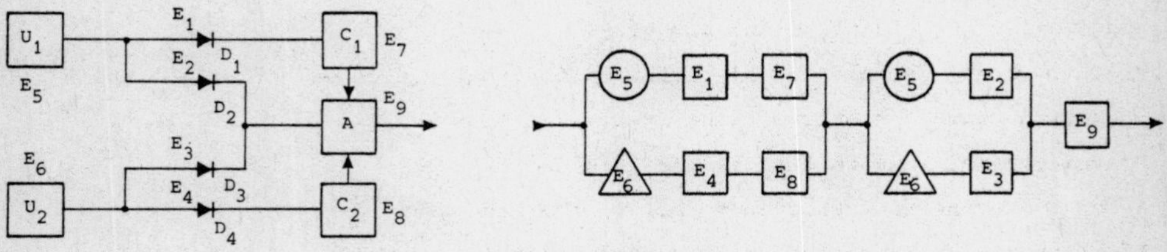

Fig. 31. **Functional and reliability block diagrams for a piece of electronic equipment with redundancy on the power supply and control units**

The required function is satisfied if, at each time point, the subassembly A and at least one of the power supplies U_1, U_2, as well as one of the control units C_1, C_2, works correctly. Considering Fig.31a and splitting up the required function as *C_1 or C_2 works correctly* and *A works correctly*, one obtains the reliability block diagram given in Fig.31b.

The reliability function $R_S(t)$ for the non-repairable case can be computed using the theorem of total probability (eq.(18)) applied to E_5 and E_6 in a probability tree. By omitting the time dependence for simplicity, and by assuming $R_5 = R_6 = R_U$, $R_1 = R_2 = R_3 = R_4 = R_D$, $R_7 = R_8 = R_C$, and $R_9 = R_A$, it follows that

$$R_S = R_U R_A [R_U (2R_D R_C - R_D^2 R_C^2)(2R_D - R_D^2) + 2(1-R_U) R_D^2 R_C]. \tag{294}$$

With $R_x = PA_x(t)$, equation (294) also gives the point-availability $PA_S(t)$ for the case in which assumption 8 above holds. However, for more general investigations of the repairable case (sections 5.2 to 5.4), the use of a computer program becomes necessary because of the large number of involved states.

6.2 Influence of preventive maintenance

Preventive maintenance is necessary to avoid wear-out failures and to identify and repair hidden failures (i.e. failures of redundant elements which can not be detected during normal operation). This section investigates some basic situations.

6.2.1 One-item repairable structures

Let us first consider a one-item repairable structure for which preventive maintenance is performed at periodic time intervals T_{PM}. The analyses are based on the following assumptions:

1. The item is new at $t = 0$.
2. Failure-free and repair times have arbitrary distribution functions, say $F(x)$ and $G(x)$, with densities $f(x)$ and $g(x)$.
3. Preventive maintenance can be performed on-line, i.e. does not affect the operating characteristics of the item.
4. Preventive maintenance time is negligible.
5. After a preventive maintenance action, the item is good-as-new.
6. If preventive maintenance falls during a repair, then it is omitted.

Considering that for reliability investigations the time points $0, T_{PM}, 2T_{PM}, 3T_{PM}$, ... are renewal points, one obtains for the reliability function (eq. (141), denoted here as $R_{PM}(t)$), the expression

$$R_{PM}(t) = 1-F(t), \qquad \text{for } 0 \leq t < T_{PM}$$

$$R_{PM}(t) = (1-F(T_{PM}))^n (1-F(t-nT_{PM})), \qquad \text{for } nT_{PM} \leq t < (n+1)T_{PM}, \quad n = 1,2,3,\ldots . \tag{295}$$

From equation (293), the mean time-to-failure $MTTF_{PM}$ is then given by

$$MTTF_{PM} = \int_0^{\infty} R_{PM}(t)\,dt = [\int_0^{T_{PM}} R_{PM}(t)\,dt][1+R_{PM}(T_{PM})+R_{PM}^2(T_{PM})+R_{PM}^3(T_{PM})+\ldots]$$

$$= \frac{\int_0^{T_{PM}} R_{PM}(t)\,dt}{1-R_{PM}(T_{PM})} = \frac{\int_0^{T_{PM}} (1-F(t))\,dt}{F(T_{PM})} . \qquad (296)$$

In the case of a constant failure rate (λ), i.e. for $F(x) = 1-e^{-\lambda x}$, equations (295) and (296) lead to $R_{PM}(t) = e^{-\lambda t}$ and $MTTF_{PM} = 1/\lambda = MTBF$, irrespective of T_{PM}. This result is a consequence of the *memoryless* property of the exponential distribution function (eq.(66)). The influence of F(x) on the reliability function $R_{PM}(t)$ is shown in Fig.32.

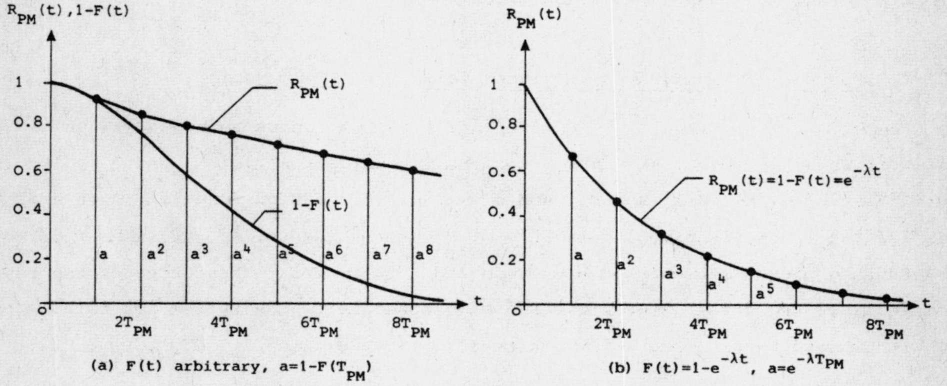

Fig. 32. **Reliability functions of a one-item structure with preventive maintenance (period T_{PM}) for two distribution functions ($F(t)$) of the failure-free times**

Computation of the point-availability (eq.(151), denoted here as $PA_{PM}(t)$), is easy if assumption 6 above can be omitted, i.e. if preventive maintenance is performed at each time point $T_{PM}, 2T_{PM}, 3T_{PM}, \ldots$, irrespective of the state of the item. In this case one obtains

$$PA_{PM}(t) = PA_u(t), \qquad \text{for } 0 \le t < T_{PM}$$
$$PA_{PM}(t) = PA_u(t-nT_{PM}), \qquad \text{for } nT_{PM} \le t < (n+1)T_{PM}, \quad n = 1,2,3,\ldots, \qquad (297)$$

with $PA_u(t)$ as in equation (151). This situation arises in practical applications when the repair times contain a large amount of logistic time. Fig.33 shows a time schedule of the function $PA_{PM}(t)$ as given by equation (297).

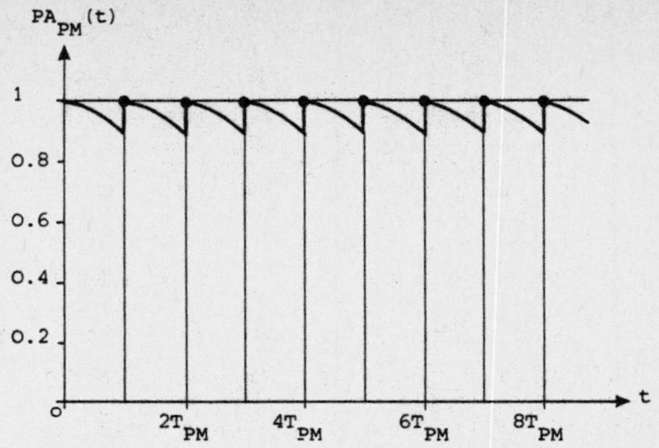

Fig. 33. **Point-availability of a one-item structure with preventive maintenance (period T_{PM}) and repair at each failure**

6.2.2 1-out-of-2 redundancy with hidden failures

As a second example let us now consider a 1-out-of-2 active redundancy with *hidden failures in one element*, say in E_1. For the investigation, let λ_1 and λ_2 be the failure rates of elements E_1 and E_2, and $G(x)$ the distribution function (with density $g(x)$) of the E_2 repair times. A failure of E_1 can be identified, and thus repaired, only during preventive maintenance or during repair of E_2. The corresponding repair time will be neglected. Fig.34 gives a time schedule of the system for both cases a) without and b) with preventive maintenance at the times $T_{PM}, 2T_{PM}, 3T_{PM}, \ldots$.

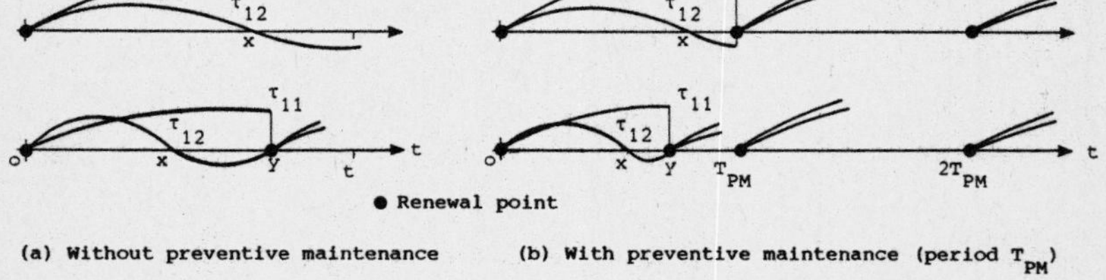

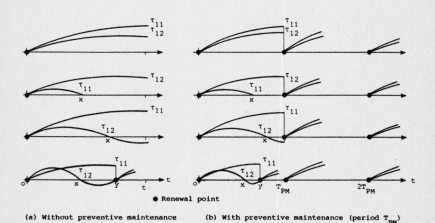

Fig. 34. **Time schedule of a repairable 1-out-of-2 active redundancy with hidden failures in element E_1**

If *no preventive maintenance is performed* and if elements E_1 and E_2 are new at $t=0$, Fig.34a leads to the following integral equation for the reliability function

$R_{SO}(t)$ (eq.(181)):

$$R_{SO}(t) = e^{-(\lambda_1+\lambda_2)t} + \int_0^t \lambda_1 e^{-\lambda_1 x} e^{-\lambda_2 t} dx + \int_0^t \lambda_2 e^{-\lambda_2 x} e^{-\lambda_1 t}(1-G(t-x))dx$$
$$+ \int_0^t \int_0^y \lambda_2 e^{-\lambda_2 x} e^{-\lambda_1 y} g(y-x) R_{SO}(t-y)\,dx\,dy. \qquad (298)$$

The Laplace transform (eq.(28)) of $R_{SO}(t)$ is then given by

$$\tilde{R}_{SO}(s) = \frac{(s+\lambda_1)(s+\lambda_1+\lambda_2)+\lambda_2(s+\lambda_2)(1-\tilde{g}(s+\lambda_1))}{(s+\lambda_1)(s+\lambda_2)(s+\lambda_1+\lambda_2)-(s+\lambda_1)(s+\lambda_2)\lambda_2\tilde{g}(s+\lambda_1)}, \qquad (299)$$

and the mean time-to-system-failure $MTTF_{SO}$ by (eq.(30))

$$MTTF_{SO} = \frac{\lambda_1(\lambda_1+\lambda_2)+\lambda_2^2(1-\tilde{g}(\lambda_1))}{\lambda_1\lambda_2(\lambda_1+\lambda_2)-\lambda_1\lambda_2^2\tilde{g}(\lambda_1)}. \qquad (300)$$

With MTTR as in equation (222) and for $\lambda_1 MTTR \ll 1$, it follows that $\tilde{g}(\lambda_1) \approx 1-\lambda_1 MTTR$ and thus

$$MTTF_{SO} \approx \frac{\lambda_1+\lambda_2+\lambda_2^2 MTTR}{\lambda_1\lambda_2(1+\lambda_2 MTTR)}. \qquad (301)$$

Assuming further that $\lambda_2 MTTR \ll 1$, equation (301) leads to

$$MTTF_{SO} \approx \frac{\lambda_1+\lambda_2}{\lambda_1\lambda_2}. \qquad (302)$$

$(\lambda_1+\lambda_2)/\lambda_1\lambda_2$ is the mean time-to-system-failure of a non-repairable standby 1-out-of-2 redundancy with failure rates λ_1 and λ_2.

Assume now that *preventive maintenance is peformed* at the times $T_{PM}, 2T_{PM}, 3T_{PM}, \ldots$, and is of negligible length. Furthermore, if the preventive maintenance falls during a repair of element E_2, then the repair is completed with the preventive maintenance in a negligible time span. As in section 6.2.1, such a situation arises in practical applications when the repair times contain a large amount of logistic time. In this case, each preventive maintenance action is a renewal point (Fig.34b). From equation (295) it follows for the reliability function $R_{SO_{PM}}(t)$ that

$$R_{SO_{PM}}(t) = R_{SO}(t), \qquad \text{for } 0 \leq t < T_{PM}$$
$$R_{SO_{PM}}(t) = R^n_{SO}(T_{PM}) R_{SO}(t-nT_{PM}), \qquad \text{for } nT_{PM} \leq t < (n+1)T_{PM}, \quad i = 1,2,3,\ldots, \qquad (303)$$

with $R_{SO}(t)$ given by equation (298). The mean time-to-system-failure $MTTF_{SO_{PM}}$ is then given by (eq.(296))

$$MTTF_{SO_{PM}} = \frac{\int_0^{T_{PM}} R_{SO}(t)\,dt}{1-R_{SO}(T_{PM})}. \qquad (304)$$

To investigate the quantity $MTTF_{SO_{PM}}$, let us assume that the repair times of element E_2 are exponentially distributed with density $g(x) = \mu e^{-\mu x}$ and thus with mean

$$MTTR = 1/\mu. \tag{305}$$

From equation (299) it follows then that

$$\tilde{R}_{SO}(s) = \frac{(s+\lambda_1+\lambda_2)(s+\lambda_1+\mu)+\lambda_2(s+\lambda_2)}{(s+\lambda_1)(s+\lambda_2)(s+\lambda_1+\lambda_2+\mu)}, \tag{306}$$

which, for $(\lambda_1+\lambda_2) \ll \mu$, leads to

$$R_{SO}(t) \simeq \frac{\lambda_2 e^{-\lambda_1 t} - \lambda_1 e^{-\lambda_2 t}}{\lambda_2 - \lambda_1}. \tag{307}$$

Putting $R_{SO}(t)$ from equation (307) into equation (304), it follows that

$$MTTF_{SO_{PM}} \simeq \frac{\frac{\lambda_2}{\lambda_1}(1-e^{-\lambda_1 T_{PM}}) - \frac{\lambda_1}{\lambda_2}(1-e^{-\lambda_2 T_{PM}})}{\lambda_2(1-e^{-\lambda_1 T_{PM}}) - \lambda_1(1-e^{-\lambda_2 T_{PM}})}; \tag{308}$$

and finally, with $e^{-\lambda x} \simeq 1-\lambda x+\lambda^2 x^2/2$,

$$MTTF_{SO_{PM}} \simeq \frac{2}{\lambda_1 \lambda_2 T_{PM}}. \tag{309}$$

A comparison between equations (302) and (309) shows the *advantage brought by preventive maintenance* to the mean time-to-system failure (factor $1/\lambda T_{PM}$ for $\lambda_1 = \lambda_2 = \lambda$). Equation (309) can be used to optimize the value of T_{PM}, obviously also handling aspects of cost and logistics.

For the investigation of other models dealing with preventive maintenance, one can refer to [44,232-235,244,253,254,256-261,281,283,299,310,312,313,325-327,330,332, 333,335,345,355,363].

6.3 Influence of imperfect switching

The influence of imperfect switching has been considered in [250,262,295-297,301,315, 328,329,336,353,354,359,362]. A first approach consists of putting the switch in series with the redundancy in the reliability block diagram. This leads to models similar to those of Fig.19 and Fig.30. For many practical applications such an approach is not appropriate, and it is necessary to *separately consider* single switches, measurement points, and control devices. Fig.35 gives such a configuration for the case of a 1-out-of-2 redundancy. From a reliability point of view, switch S_i, element

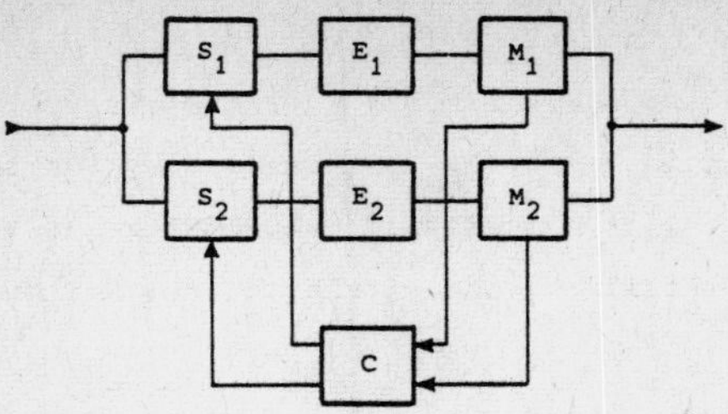

Fig. 35. **Functional block diagram for a 1-out-of-2 redundancy with switches (S_1,S_2), measurement points (M_1,M_2), and control device (C)**

E_i, and measurement point M_i in Fig.35 are in series ($i = 1,2$). To simplify the investigations, let τ_{b1}, τ_{b2} and τ_c be, respectively, the failure-free times of the group (S_1,E_1,M_1), of the group (S_2,E_2,M_2), and of the control device C. The corresponding distribution functions are $F_b(x)$ for τ_{b1} and τ_{b2} and $F_c(x)$ for τ_c, with densities $f_b(x)$ and $f_c(x)$. Repair is not considered.

Assuming first that at $t=0$, element E_1 is put into the operating state and element E_2 into the *standby mode*, a system failure in the time interval (0,t) occurs with one of the following mutually exclusive events:

- $\tau_c > \tau_{b1} \cap (\tau_{b1}+\tau_{b2}) \leq t$
- $\tau_c < \tau_{b1} \leq t$.

It is therefore implicity assumed that a failure of the control device has no influence on the operating element and does not consist of a commutation to E_2. With these assumptions, the reliability function $R_S(t) = \Pr\{\text{system up in } (0,t)\}$ of the system described by Fig.35 is given by

$$R_S(t) = 1-[\int_0^t f_b(x)(1-F_c(x))F_b(t-x)dx+\int_0^t f_b(x)F_c(x)dx]. \tag{310}$$

For $f_b(x) = \lambda_b e^{-\lambda_b x}$ and $f_c(x) = \lambda_c e^{-\lambda_c x}$ it follows that

$$R_S(t) = e^{-\lambda_b t}+(1-e^{-\lambda_c t})\frac{\lambda_b}{\lambda_c}e^{-\lambda_b t}, \tag{311}$$

the corresponding mean time-to-system failure $MTTF_S$ being (eq.(27))

$$MTTF_S = \frac{2\lambda_b+\lambda_c}{\lambda_b(\lambda_b+\lambda_c)}. \tag{312}$$

Finally, for $\lambda_c \equiv 0$, i.e. for a 100% reliable control device, one obtains from equations (311) and (312) the results for a non-repairable 1-out-of-2 standby redundancy (eq.(32) and (33) with $n=2$, $k=1$, $\lambda=\lambda_b$).

Assume now that at $t=0$, element E_1 is put into the operating state and element E_2 into the *active redundancy mode*. In this case, a system failure in the time interval $(0,t)$ occurs with one of the following mutually exclusive events (see above for the assumptions about the failure mode of the control device):

- $\tau_{b1} \leq t \cap \tau_c > \tau_{b1} \cap \tau_{b2} \leq t$
- $\tau_c < \tau_{b1} \leq t$.

From this, the reliability function $R_S(t)$ of the system described by Fig.35 is given by

$$R_S(t) = 1-[F_b(t)\int_0^t f_b(x)(1-F_c(x))dx+\int_0^t f_b(x)F_c(x)dx]. \tag{313}$$

The case of constant failure rates (λ_b,λ_c) then leads to

$$R_S(t) = \frac{2\lambda_b+\lambda_c}{\lambda_b+\lambda_c}\, e^{-\lambda_b t} - \frac{\lambda_b}{\lambda_b+\lambda_c}\, e^{-(2\lambda_b+\lambda_c)t}, \tag{314}$$

and to a mean time-to-system-failure $MTTF_S$ given by

$$MTTF_S = \frac{2\lambda_b+\lambda_c}{\lambda_b(\lambda_b+\lambda_c)} - \frac{\lambda_b}{(\lambda_b+\lambda_c)(2\lambda_b+\lambda_c)} . \tag{315}$$

For $\lambda_c \equiv 0$ one obtains from equations (314) and (315) the results for 1-out-of-2 active redundancy (eq.(17) or (29)).

REFERENCES

Books, monographs, general papers

[1] R.T. Anderson et al., Reliability design handbook, RDH-376, New York: Reliability Analysis Center/Griffis Air Force Base, 1976.

[2] A.M. Andronov et al., "Statistic of semi-Markov birth and death processes with application to queuing systems", Eng.Cybernetics, no.3, pp.465-472, 1972.

[3] R.E. Barlow et al., Mathematical theory of reliability, New York: Wiley, 1965; Statistical theory of reliability and life testing, New York: Holt,Rinehart & Winston, 1975.

[4] M.S. Barlett, Stochastic processes, Cambridge: University Press, 1978.

[5] Yu.K. Belyayev et al., "On some stochastic problems of reliability theory", Proc. of the 5th Berkeley Symp. on Math. Stat. and Prob., vol.3, pp.259-270, 1965.

[6] F.J. Beutler et al., "The theory of stationary point processes", Acta Mathematica, pp.159-197, 1966.

[7] A. Birolini, "Some applications of regenerative stochastic processes to reliability theory", IEEE Trans.Rel., vol.R-23, pp.186-194, 1974 and Vol.R-24, pp.336-340, 1975; "Hardware simulation of semi-Markov and related processes", Math. & Comp. in Simulation, vol.19, pp.75-97 and pp.183-191, 1977 (also in the PhD-Thesis no.5375, ETH Zurich, 1974).

[8] A. Birolini, Zuverlässigkeit von Schaltungen und Systemen, ETH Zurich, 4th ed.1982.

[9] A. Birolini, Qualität und Zuverlässigkeit technischer Systeme: Theorie, Praxis, Management, Berlin/Heidelberg/New York/Tokyo: Springer-Verlag,1985.

[10] P. Bitter et al., Technische Zuverlässigkeit, Berlin/Heidelberg/New York: Springer-Verlag, 1972.

[11] Centre National d'Etudes des Télécommunications, Recueil de données de fiabilité du CNET, Lannion: Centre National d'Etudes des Télécommunications, 1983.

[12] E. Çinlar, Introduction to stochastic processes, Englewood Cliffs, N.J.: Prentice-Hall,1975.

[13] D.R. Cox, "The analysis of non-Markovian stochastic processes by the inclusion of supplementary variables", Proc.Cambridge Phil.Soc., vol.51, pp.433-441, 1955.

[14] D.R. Cox, Renewal theory, London: Methuen, 1962.

[15] H. Cramér, "Model building with the aid of stochastic processes", Technometrics, vol.6, no.2, pp.133-159, May 1964.

[16] M.TS. Dimitrov, "A limit theorem for a duplicated system with unrenewable redundancy", Eng.Cybernetics, no.10, pp.816-819, 1972.

[17] I.L. Doob, Stochastic processes, New York: Wiley, 1967.

[18] W. Feller, "On the integral equation of renewal theory", Ann.Math.Statistics, vol.12, pp.243-267, 1941.

[19] W. Feller, An introduction to probability theory and its applications, New York: Wiley, vol.I 1957, vol.II 1966.

[20] P. Franken et al., "Estimates of the indices of reliability for redundant systems with renewal", Eng.Cybernetics, no.4, pp.64-69, 1977.

[21] K.W. Gaede, Zuverlässigkeit Mathematische Modelle, München: Carl Hanser Verlag, 1977.

[22] Ya.G. Genis, "The asymptotic behaviour of a flow of rare events in a regenerating process", Eng.Cybernetics, no.5, pp.77-84, 1978; "The consequence of a certain class of distribution functions to an exponential distribution function in reliability and queueing problems", Eng.Cybernetics, no.6, pp.94-101, 1978.

[23] B.V. Gnedenko, "Idle duplication", Eng.Cybernetics, no.4, pp.1-9, 1964.

[24] B.V. Gnedenko, "Duplication with repair", Eng.Cybernetics, no.5, pp.102-108, 1964.

[25] B.V. Gnedenko, "Some theorems on standbys", in Proc. 5th Berkeley Symp.Math.Stat.Probability, vol.3, pp.285-291, 1965.

[26] B.V. Gnedenko et al., Mathematical methods of reliability theory, New York: Academic, 1969. Moscow: Nauka, 1965. Berlin: Akademic, 1968.

[27] B.V. Gnedenko et al., "A general model for standby with renewal", Eng.Cybernetics, no.6, pp.82-86, 1974.

[28] B.I. Grigelionis, "Limit theorems for sums of repair processes", Cybernetics in the Service of Communism, no.2, pp.316-341, 1964.

[29] U. Höfle-Isphording, Zuverlässigkeitsrechnung, Berlin/Heidelberg/New York: Springer-Verlag, 197

[30] IEEE Trans.Rel., Special Issue on: Nuclear System Reliability and Safety, IEEE Trans.Rel., vol.R-25, Aug.1976.

[31] J. Juran et al., Quality control handbook, New York: Mc.Graw-Hill, 1974.

[32] Kh.K. Karapenev, "Estimation of the reliability of a standby system with renewal and lag in the detection of failures", Eng.Cybernetics, no.3, pp.88-93, 1978

[33] S. Karlin et al., "The differential equations of birth and death processes, and the Stieltjes moment problem", Trans.Amer.Math.Soc., vol.85, pp.489-546, 1957; "The classification of birth and death processes", Trans.Amer.Math.Soc., vol.86, pp.366-400, 1957; "Coincidence properties of birth and death processes", Pacific J.Math., vol.9, pp.1109-1140, 1959.

[34] A. Kaufmann et al., Mathematical models for the study of the reliability systems, New York: Academic Press, 1977.

[35] Z. Khalil, "Asymptotic distribution of a standby system with delayed repair", IEEE Trans. Rel., vol.R-28, pp.265-266, Aug.1979.

[36] A.Ja. Khintchine, Desserte d'un grand nombre d'usagers, Paris: Dunod, 1969.

[37] B.A. Kozlov et al., Reliability handbook, New York: Holt, Rinehart & Winston, 1970.

[38] P. Lévy, "Processus semi-Markoviens", Proc. of the Int. Congress of Math., Amsterdam, vol.3, pp.416-426, 1954.

[39] J. Mayer, "Ausfallprozesse: Mathematische Modelle und Zusammenhänge", Sonderdruck aus den Siemens-Albis Berichten, no.31,32,33, 41p., 1982.

[40] MIL-HDBK-217, Reliability Prediction of Electronic Equipment, USA Department of Defence, Edition D, 1982, not.1, 1983.

[41] MIL-STD-785, Reliability Program for System and Equipment Development and Production, USA Department of Defence, Edition B, 1980.

[42] MIL-STD-1629, Procedures for Performing a Failure Mode and Effects Analysis for Shipboard Equipment, USA Department of Defence, Edition A, 1980.

[43] S. Osaki et al., "Bibliography of reliability and availability of stochastic systems", IEEE Trans.Rel., Vol.R-25, pp.284-287, Oct.1976.

[44] S. Osaki et al. (eds.), Stochastic models in reliability theory, Lecture Notes in Economics and Math.Systems, no.235, Berlin/Heidelberg/New York/Tokyo: Springer-Verlag, 1984.

[45] E. Parzen, Stochastic processes, San Francisco: Holden-Day, 1962.

[46] I.V. Pavlov et al., "The asymptotic distribution of the time until a semi-Markov process gets out of a kernel", Eng.Cybernetics, no.5, pp.68-72, 1978.

[47] R. Pyke, "Markov renewal processes: definitions and preliminary properties", Ann.Math.Stat., vol.32, pp.1231-1242, 1961; "Markov renewal processes with finitely many states", Ann.Math. Stat., vol.32, pp.1243-1259, 1961; R. Pyke et al., "Limit theorems for Markov renewal processes", Ann.Math.Stat., vol.35, pp.1746-1764, 1964; "The existence and uniqueness of stationary measures for Markov renewal processes", Ann.Math.Stat., vol.37, pp.1439-1462, 1966.

[48] K. Reinschke, Zuverlässigkeit von Systemen, Band I: Systeme mit endlich vielen Zuständen, Berlin: VEB Verlag Technik, 1973.

[49] W. Schneeweiss, Zuverlässigkeitstheorie, Berlin/Heidelberg/New York: Springer-Verlag, 1973.

[50] M. Schwob et al., Traité de fiabilité, Paris: Masson, 1969.

[51] A.A. Shakhbazov, "Estimation of the reliability of a complex system with rapid renewal in a nonstationary regime", Eng.Cybernetics, no.1, pp.58-66, 1981; "Limiting distribution of the first entry for semi-Markov processes and its application to reliability theory", Eng.Cybernetics, no.3, pp.82-90, 1982.

[52] M.L. Shooman, Probabilistic reliability: an engineering approach, New York: Mc.Graw-Hill, 1968.

[53] W.L. Smith, "Asymptotic renewal theorems", Proc.Roy.Soc., Edimbourgh, vol.64, pp.9-48,1954.

[54] W.L. Smith, "Renewal theory and its ramifications", J.Royal Stat.Soc., Ser.B, vol.20, pp.243-302, 1958.

[55] W.L. Smith, "Regenerative stochastic processes", in Proc.Int.Congr.Math., vol.3, pp.304-305, 1954; "Regenerative stochastic processes", Proc.Royal Soc., London, Ser.A, vol.232, pp.6-31, 1955; "Remarks on the paper: Regenerative stochastic processes", Proc.Royal Soc., London, Ser.A., vol.256, pp.496-501, 1960.

[56] A.D. Solovyev, "Asymptotic distribution of lifetime of a duplicated elements", Eng.Cybernetics, no.5, pp.109-11, 1964.

[57] A.D. Solovyev, "Standby with rapid renewal", Eng.Cybernetics, no.1, pp.49-64, 1970; "Asymptotic behaviour of the time of first occurrence of a rare event", Eng.Cybernetics, no.6, pp.1038-1048, 1971.

[58] A.D. Solovyev et al., "Standby with incomplete renewal", Eng.Cybernetics, no.5, pp.58-62,1975.

[59] A.D. Solovyev et al., "Nonhomogeneous standby with renewal", Eng.Cybernetics, no.5, pp.30-38, 1981.

[60] S.K. Srinivasan, Stochastic point processes and their applications, London: Griffin, 1974.

[61] S.K. Srinivasan et al., Probabilistic analysis of redundant systems, Lecture Notes in Economics and Math.Systems, no.175, Berlin/Heidelberg/New York: Springer-Verlag, 1980.

[62] H. Störmer, Mathematische Theorie der Zuverlässigkeit, München: Oldenbourg, 1970.

[63] H. Störmer, Semi-Markoff-Prozesse mit endlich vielen Zuständen, Lecture Notes in Operations Research and Math. Systems, no.34,Berlin/Heidelberg/New York: Springer-Verlag, 1970.

[64] L. Takács, "On certain sojourn time problems in the theory of stochastic processes", Acta Mathematica, Hungar, vol.8, pp.169-191, 1957.

[65] L. Takács, Stochastic processes: problems and solutions, London: Methuen & Science Paperback, USA: Barnes & Noble, 1966.

[66] US Army Material Command, Development guide for reliability, Part 1 to 6, AMC Pamphlet 706-195...706-200, Alexandria/Va, 1975/76.

[67] V.A. Zaytsev et al.,"Redundancy of complex systems", Eng.Cybernetics, no.4, pp.66-75, 1975; V.A. Zaytsev, "The speed of convergence of the probability of failure-free operation to an exponential distribution for duplicated systems with renewals", Eng.Cybernetics, no.2, pp. 63-69, 1976.

One-item repairable structures

[68] K. Adachi et al., "Optimum policy on one-unit system with two types of maintenance and minimal repair", Microel. & Rel., vol.20, pp.489-493, 1980.

[69] R. Barlow et al., "Reliability analysis of a one-unit system", Op.Research, vol.9, pp.200-208, 1961.

[70] L.A. Baxter, "Availability measures for a two-state system", J.Appl.Prob., vol.18, pp.227-235, 1981.

[71] A. Birolini, "Generalization of the expressions for the reliabilities and availabilities of a repairable item", Proc. 2nd Int.Conf. on Structural Mechanics in Reactor Technology, contribution M8/3, 16p., Berlin, Sept. 10-14, 1973.

[72] S. Garribba et al., "Availability of repairable units when failure and restoration rates age in real time", IEEE Trans.Rel., vol.R-25, pp.88-94, June 1976.

[73] J.E. Hosford, "Measures of dependability", Oper.Res., vol.8, pp.53-64, 1960.

[74] M. Kodama et al., "Mission reliability for a 1-unit system with allowed down-time", IEEE Trans.Rel., vol.R-22, pp.268-270, Dec.1973.

[75] T.J. Li, "On the calculation of system downtime distribution", IEEE Trans.Rel., vol.R-20, pp.38-39, Febr.1971.

[76] H.F. Martz, Jr., "On single-cycle availability", IEEE Trans.Rel., vol.R-20, pp.21-23, Febr. 1971.

[77] H. Mine et al., "Preventive replacement of a 1-unit system with a wearout state", IEEE Trans.Rel., vol.R-23, pp.24-29, Apr.1974.

[78] H. Mine et al., "Optimal ordering and replacement for 1-unit system", IEEE Trans.Rel., vol.R-26, pp.273-276, Oct.1977.

[79] E.J. Muth, "A method for predicting system downtime", IEEE Trans.Rel., vol.R-17, pp.97-102, June 1968.

[80] E.J. Muth, "Excess time, a measure of system repairability", IEEE Trans.Rel., vol.R-19, pp.16-19, Febr.1970.

[81] T. Nakagawa et al., "A note on availability for a finite interval", IEEE Trans.Rel., vol.R-22, pp.271-272, Dec.1973.

[82] T. Nakagawa et al., "Off time distributions in an alternating renewal process with reliability applications", Microel. & Rel., vol.13, pp.181-188, 1974.

[83] M. Sasaki et al., "Availabilities for a fixed periodic system", IEEE Trans.Rel., vol.R-26, pp.300-302, Oct.1977.

[84] F.A. Tillman et al., "Numerical evaluation of instantaneous availability", IEEE Trans.Rel., vol.R-32, pp.119-123, Apr.1983.

[85] S. Yamada et al., "Checking request policies for a one-unit system and their comparisons", Microel. & Rel., vol.20, pp.859-874, 1980.

Two-item repairable redundancies

[86] J.R. Arora, "Reliability of a 2-unit standby redundant systems with constrained repair time", IEEE Trans.Rel., vol.R-25, pp.203-205, Aug.1976.

[87] J.R. Arora, "Reliability of a 2-unit priority-standby redundant system with finite repair capability", IEEE Trans.Rel., vol.R-25, pp.205-207, Aug.1976.

[88] K. Arndt, "Calculation of the reliability of a duplicated system by the method of embedded semi-Markov processes", Eng.Cybernetics, no.3, pp.54-63, 1977.

[89] A. Birolini, "Comments on: renewal theoretic aspects of two-unit redundant systems", IEEE Trans.Rel., vol.R-21, pp.122-123, May 1972.

[90] A. Birolini, "Some applications of regenerative stochastic processes to reliability theory, Part II: Reliability and availability of 2-item redundant systems", IEEE Trans.Rel., vol.R-24, pp.336-340, Dec.1975.

[91] M.H. Branson et al., "Reliability analysis of systems comprised of units with arbitrary repair-time distributions", IEEE Trans.Rel., vol.R-20, pp.217-223, Nov.1971.

[92] J.A. Buzacott, "Availability of priority standby redundant systems", IEEE Trans.Rel., vol.R-20, pp.60-63, May 1971.

[93] J.A. Buzacott, "Reliability analysis of a nuclear reactor fuel charging system", IEEE Trans.Rel., vol.R-22, pp.88-91, June 1973.

[94] D.K. Chow, "Reliability of some redundant systems with repair", IEEE Trans.Rel., vol.R-22, pp.223-228, Oct.1973.

[95] B. Epstein et al., "Reliability of some two unit redundant systems", Proc.Annu.Symp.Reliability, pp.469-476, 1960.

[96] J. Fukuta et al., "Mission reliability for a redundant repairable system with two dissimilar units", IEEE Trans.Rel., vol.R-23, pp.280-282, Oct.1974.

[97] R.C. Garg et al., "Reliability prediction of a two-unit standby redundant system with standby failure", Microel. & Rel., vol.11, pp.263-267, 1972.

[98] D.P. Gaver, Jr., "Time to failure and availability of paralleled systems with repair", IEEE Trans.Rel., vol.R-12, pp.30-38, June 1963.

[99] D.P. Gaver, Jr., "Failure time for a redundant repairable system of two dissimilar elements", IEEE Trans.Rel., vol.R-13, pp.14-22, Mars 1964.

[100] M.N. Gopalan et al., "s-Expected number of repairs from online, standby and frequency of failure of a 1-server 2-unit warm standby system", Microel. & Rel., vol.22, pp.45-51, 1982.

[101] M.N. Gopalan et al., "Cost-benefit analysis of a 1-server 2-unit system subject to different repair strategies", Microel. & Rel., vol.22, pp.393-397, 1982.

[102] M.N. Gopalan et al., "s-Expected busy period of 1-server 2-unit warm standby system subject to different repair strategies", Microel. & Rel., vol.22, pp.399-403, 1982.

[103] M.N. Gopalan et al., "Stochastic behaviour of a two-unit repairable system subject to inspection", Microel. & Rel., vol.22, pp.717-722, 1982.

[104] R. Harris, "Reliability applications of a bivariate exponential distribution", Oper.Res., vol.16, pp.18-27, 1968.

[105] D.V.S. Kapil et al., "Repair limit suspension policies for a 2-unit redundant system with 2-phase repairs", IEEE Trans.Rel., vol.R-30, p.90, Apr.1981.

[106] K.R. Kapoor et al., "Analysis of a 2-unit standby redundant repairable systems, IEEE Trans. Rel., vol.R-27, pp.385-388, Dec.1978.

[107] P.K. Kapur et al., "A 2-unit warm-standby redundant system with delay and one repair facility", IEEE Trans.Rel., vol.R-24, p.275, Oct.1975.

[108] P.K. Kapur et al., "Stochastic behaviour of some 2-unit redundant systems", IEEE Trans.Rel., vol.R-27, pp.382-385, Dec.1978.

[109] P.K. Kapur et al., "Interval reliability of a two-unit stand-by redundant system", Microel. & Rel., vol.23, pp.167-168, 1983.

[110] M. Kodama et al., "Reliability considerations for a 2-unit redundant system with Erlang-failure and general repair distributions", IEEE Trans.Rel., vol.R-23, pp.75-81, June 1974.

[111] M. Kodama, "Reliability analysis of a 2-dissimilar units redundant system with Erlang-failure and general repair distributions", Microel. & Rel., vol.13, pp.523-528, 1974.

[112] M. Kodama et al., "Analysis of 7 models for the 2-dissimilar-unit, warm standby, redundant system, IEEE Trans.Rel., vol.R-25, pp.273-279, Oct.1976.

[113] D.K. Kulshrestha, "Reliability of a parallel redundant complex system", Oper.Res., vol.16, pp.28-35, 1968.

[114] D.K. Kulshrestha, "Application of discrete transforms in reliability of a parallel redundant system", Metrika, vol.18, pp.145-153, 1972.

[115] A. Kumar et al., "On a 2-unit standby system with repair specification", IEEE Trans.Rel., vol.R-26, pp.369-370, Dec.1977.

[116] A. Kumar et al., "Optimal maintenance of a two-unit standby redundant system with a generalized cost structure", Microel. & Rel., vol.21, pp.117-120, 1981.

[117] R. Lal et al., "2-unit standby system with fault analysis", IEEE Trans.Rel., vol.R-29, pp.431-432, Dec.1980.

[118] B.H. Liebowitz, "Reliability considerations for a two element redundant systems with generalized repair times", Oper.Res., vol.14, pp.233-241, 1966.

[119] D.G. Linton et al., "Laplace transforms for the two-unit cold-standby redundant system", IEEE Trans.Rel., vol.R-22, pp.105-108, June 1973.

[120] D.G. Linton, "Some advancements in the analysis of two-unit parallel redundant systems", Microel. & Rel., vol.15, pp.39-46, 1976.

[121] H. Mine et al., "Reliability considerations on redundant systems with repair, Mem.Fac.Eng. Kyoto Univ., vol.29, pp.509-529, 1967.

[122] H. Mine et al., "On failure time distributions for systems of dissimilar units", IEEE Trans. Rel., vol.R-18, pp.165-168, Nov.1969.

[123] H. Mine et al., "The effect of an age replacement to a standby redundant system", J.Appl. Prob., vol.6, pp.516-523, Dec.1969.

[124] H. Mine et al., "Repair priority effect on availability of a 2-unit system" IEEE Trans.Rel., vol.R-28, pp.325-326, Oct.1979.

[125] K. Murari et al., "2-unit standby system with proviso for rest and maximum rest", Microel. & Rel., vol.21, pp.573-579, 1981.

[126] E.J. Muth, "Reliability of a system having a standby spare plus multiple repair capability", IEEE Trans.Rel., vol.R-15, pp.76-81, Aug.1966.

[127] T. Nakagawa et al., "Stochastic behaviour of a two-unit standby redundant system", INFOR, vol.12, pp.66-70, Feb.1974.

[128] T. Nakagawa et al., "Stochastic behaviour of a two-unit priority standby redundant system with repair", Microel. & Rel., vol.14, pp.309-313, 1975.

[129] T. Nakagawa et al., "Stochastic behaviour of two-unit paralleled redundant systems with repair maintenance", Microel. & Rel., vol.14, pp.457-461, 1975.

[130] M. Ohashi et al., "Stochastic behaviour of a two-unit parallel system", Mircoel. & Rel., vol.20, pp.471-476, 1980.

[131] M. Ohashi et al., "A two-unit paralleled system with general distribution", J. of the Oper. Res., vol.23, pp.313-323, Dec.1980.

[132] K. Okumoto, "Availability of a 2-component dependent system", IEEE Trans.Rel., vol.R-30, p.205, June 1981.

[133] S. Osaki, "Reliability analysis of two-unit standby redundant systems", Proc. 3th Hawaii Int.Conf. on System sciences, pp.53-56, 1970.

[134] S. Osaki, "Reliability analysis of a two-unit standby redundant system with priority", Canadian Oper.Res.Soc.J., vol.8, pp.60-62, Mars 1970.

[135] S. Osaki, "Renewal theoretic aspects of two-unit redundant systems", IEEE Trans.Rel., vol.R-19, pp.105-110, Aug.1970.

[136] S. Osaki et al., "On a two-unit standby redundant system with standby failure", Oper.Res., vol.19, pp.510-523, 1971.

[137] S. Osaki, "A two-unit parallel redundant system with bivariate exponential lifetime", Microel. & Rel., vol.20, pp.521-523, 1980.

[138] K.G. Ramamurthy et al., "A two-dissimilar-unit cold standby system with allowed down time", Microel. & Rel., vol.22, pp.689-691, 1982.

[139] R. Ramanarayanan et al., "A 2-unit cold standby system with Marshall-Olkin bivariate exponential life andrepair times" IEEE Trans.Rel., vol.R-30, pp.489-490, Dec.1981.

[140] N. Ravichandran, "Analysis of two-unit parallel redundant system with phase type failure and general repair", Microel. & Rel., vol.21, pp.569-572, 1981.

[141] N. Ravichandran, "Availability measures for a class of redundant systems", Eurocon'82, Copenhagen, June 14-18, 1982. Amsterdam: North-Holland Publ., 1982.

[142] M. Sasaki et al., "Improvement of instantaneous availabilities by decreasing delay time", IEEE Trans.Rel., vol.R-29, pp.178-181, June 1980.

[143] J. Skákala et al., "2-unit redundant systems with replacement & repair", IEEE Trans.Rel., vol.R-26, pp.294-296, Oct.1977.

[144] S.K. Srinivasan et al., "Probabilistic analysis of a two-unit system with a warm standby and a single repair facility", Oper.Res., vol.21, pp.748-754, 1973.

[145] S.K. Srinivasan et al., "Probabilistic analysis of a 2-unit cold-standby system with a single repair facility", IEEE Trans.Rel., vol.R-22, pp.250-254, Dec.1973.

[146] S.K. Srinivasan et al., "Analysis of intermittently used 2-unit redundant systems with a single repair facility", Microel. & Rel., vol.19, pp.247-252, 1979.

[147] V.S. Srinivasan, "The effect of standby redundancy in system's failure with repair maintenance", Oper.Res., vol.14, pp.1024-1036, 1966.

[148] A. Subramanian, "Probabilistic analysis of a system with a 2-unit subsystem", IEEE Trans. Rel., vol.R-30, p.489, Dec.1981.

[149] R. Subramanian et al., "Interval reliability of a 2-unit redundant system", IEEE Trans.Rel., vol.R-28, p.84, Apr.1979.

[150] R. Subramanian et al., "A two-unit redundant system", Microel. & Rel., vol.19, pp.277-278, 1979.

[151] R. Subramanian et al., "Probabilistic analysis of a two-unit parallel redundant system", Microel. & Rel., vol.19, pp.321-323, 1979.

[152] R. Subramanian et al., "Stochastic behaviour of 2-unit parallel-redundant system", IEEE Trans.Rel., vol.R-28, pp.419-420, Dec.1979.

[153] R. Subramanian et al., "Availability analysis of a 2-unit redundant system", IEEE Trans. Rel., vol.R-29, pp.182-183, June 1980.

[154] R. Subramanian et al., "Stochastic models of 2-unit systems", IEEE Trans.Rel., vol.R-30, pp.85-86, Apr.1981.

[155] R. Subramanian et al., "Study of a 2-unit priority-standby system", IEEE Trans.Rel., vol. R-30, pp.388-390, Oct.1981.

[156] D. Wiens, "Analysis of a hot-standby system with 2 identical, dependent units and a general Erlang failure time distribution", IEEE Trans.Rel., vol.R-30, p.386, Oct.1981.

[157] A.F. Zubova, "Idle duplication with repair for any distribution of flow of breakdowns and time of repair", Eng.Cybernetics, no.2, pp.99-111, 1964.

Complex repairable structures

[158] K. Adachi et al., "k-out-of-n:G system with simultaneous failure and three repair policies", Microel. & Rel., vol.19, pp.351-361, 1979.

[159] M. Agarwal et al., "Analysis of 2 models of standby redundant systems", IEEE Trans.Rel., vol.R-29, pp.84-85, Apr.1980.

[160] V. Amoia et al., "Computer-oriented formulation of transition-rate matrices via Kronecker algebra", IEEE Trans.Rel., vol.R-30, pp.123-132, June 1981.

[161] J.R. Arora, "Reliability of several standby-priority-redundant systems", IEEE Trans.Rel., vol.R-26, pp.290-293, Oct.1977.

[162] H.E. Ascher et al., "Repairable systems reliability: future research topics", Eurocon'82, Copenhagen, June 14-18, 1982. Amsterdam: North-Holland Publ., 1982.

[163] U.N. Bhat, "Reliability of an independent component, s-spare system with exponential life times and general repair times", Technometrics, vol.15, pp.529-539, Aug.1973.

[164] A. Birolini, "On the use of stochastic processes in modeling reliability problems", Eurocon'82, Copenhagen, June 14-18, 1982. Amsterdam: North-Holland Publ., pp.96-100, 1982.

[165] A. Bobbio et al., "Multi-state homogeneous Markov models in reliability analysis", Microel. & Rel., vol.20, pp.875-880, 1980.

[166] J.A. Buzacott, "Markov approach to finding failure times of repairable systems", IEEE Trans.Rel., vol.R-19, pp.128-134, Nov.1970 and "Network approaches to finding the reliability of repairable systems", IEEE Trans.Rel., Vol.R-19, pp.140-146, Nov.1970.

[167] D.K. Chow, "Reliability of two items in sequence with sensing and switching", IEEE Trans. Rel., vol.R-20, pp.254-256, Nov.1971.

[168] D.K. Chow, "Availability of some repairable computer systems", IEEE Trans.Rel., vol.R-24, pp.64-66, 1975.

[169] W.R. Christiaanse, "A technique for the analysis of repairable redundant systems", IEEE Trans.Rel., vol.R-19, pp.53-60, May 1970.

[170] C.A. Clarotti, "Limitations of minimal cut-set approach in evaluating reliability of systems with repairable components", IEEE Trans.Rel., vol.R-30, pp.335-338, Oct.1981.

[171] F. Downton, "The reliability of multiplex systems with repair", J.Roy.Statist.Soc., Ser.B, vol.28, pp.459-476, 1966.

[172] H. Frey, "Computerorientierte Methodik der Systemzuverlässigkeits- und Sicherheitsanalyse", PhD-Thesis no.5244, ETH Zürich, 1973.

[173] F. Galetto, "System availability and reliability analysis", Proc.Ann.Rel.& Maint.Symp., pp.95-100, 1977.

[174] D.G. Gnedenko et al., "Estimation of the reliability of complex renewable systems", Eng. Cybernetics, no.3, pp.89-96, 1975.

[175] M.N. Gopalan, "Availability and reliability of a series-parallel system with a single repair facility", IEEE Trans.Rel., vol.R-24, pp.219-220, Aug.1975.

[176] M.N. Gopalan, "Analysis of two different 1-server systems", IEEE Trans.Rel., vol.R-25, pp.279-280, Oct.1976.

[177] M.N. Gopalan et al., "Analysis of a n-unit system with 1-repair facility which has downtime", IEEE Trans.Rel., vol.R-26, pp.297-298, Oct.1977.

[178] M.N. Gopalan et al., "Availability and reliability of a 1-server system with n warm standbys", IEEE Trans.Rel., vol.R-26, p.298, Oct.1977.

[179] M.N. Gopalan et al., "Analysis of 1-server n-unit system", IEEE Trans.Rel., vol.R-29, p.187, June 1980.

[180] M.N. Gopalan et al., "s-Expected number of repairs and frequency of failures of a n-unit system with a single repair facility", Microel. & Rel., vol.21, pp.581-584, 1981.

[181] M.N. Gopalan et al., "Stochastic behaviour of a 1-server n-unit system subject to general repair distribution", Microel. & Rel., vol.21, pp.585-588, 1981.

[182] A.K. Govil, "Mean time to system failure for a 4-unit redundant repairable system", IEEE Trans.Rel., vol.R-23, pp.56-57, Apr.1974.

[183] H. Gupta et al., "A method of symbolic steady-state availability evaluation of k-out-of-n: G system", IEEE Trans.Rel., vol.R-28, pp.56-57, Apr.1979.

[184] P.P. Gupta et al., "Availability of a parallel redundant complex system", IEEE Trans.Rel., vol.R-27, pp.389-390, Dec.1978.

[185] R.A. Hall et al., "Reliability of non-exponential redundant systems", Proc.Annu.Symp.Reliability, pp.594-608, 1966.

[186] C.L. Hwang et al., "System-reliability evaluation techniques for complex/large systems-a review", IEEE Trans.Rel., vol.R-30, pp.416-423, Dec.1981.

[187] T. Ito et al., "Reliability of special redundant systems considering exchange time and repair time", IEEE Trans.Rel., vol.R-20, pp.11-16, Febr.1971.

[188] S. Kalpakam et al., "Availability of a special series system", IEEE Trans.Rel., vol.R-30, pp.202-203, June 1981.

[189] L. Kanderhag, "Eigenvalue approach for computing the reliability of Markov systems", IEEE Trans.Rel., vol.R-27, pp.337-340, Dec.1978.

[190] P.K. Kapur et al., "Intermittently used redundant systems", Microel. & Rel., vol.17, pp.593-596, 1978.

[191] M. Kodama et al., "Reliability and maintainability of a multicomponent series-parallel system under several repair disciplines", Microel. & Rel., vol.22, pp.1135-1153, 1982.

[192] D.K. Kulshrestha, "Reliability of a parallel redundant complex system", Oper.Res., vol.16, pp.28-35, 1968.

[193] D.K. Kulshrestha, "Reliability of a repairable multicomponent system with redundancy in parallel", IEEE Trans.Rel., vol.R-19, pp.50-52, May 1970.

[194] M. Kumagai, "Reliability analysis for systems with repairs", J.Oper.Res.Soc.Japan, vol.14, pp.53-71, Sept.1971.

[195] M. Kumagai, "Availability of an n-spare system with a single repair facility", IEEE Trans. Rel., vol.R-24, pp.216-217, Aug.1975.

[196] A. Kumar,"Stochastic behaviour of a special complex system", IEEE Trans.Rel., vol.R-25, pp.108-109, June 1976.

[197] A. Kumar et al., "Stochastic behaviour of a standby redundant system", IEEE Trans.Rel., vol. R-27, pp.169-170, June 1978.

[198] J.C. Laprie, "On reliability prediction of repairable redundant digital structures", IEEE Trans.Rel., vol.R-25, pp.256-258, Oct.1976.

[199] T.Y. Liang, "Availability of a special 2-unit series system", IEEE Trans.Rel., vol.R-27, pp.294-297, Oct.1978.

[200] D.G. Linton et al., "Reliability analysis of the k-out-of-n:F system", IEEE Trans.Rel., vol. R-23, pp.97-103, June 1974.

[201] D.G. Linton, "Life distributions and degradation for a 2-out-of-n:F system", IEEE Trans. Rel., vol.R-30, pp.82-84, Apr.1981.

[202] H. Mine et al., "Some considerations for multiple-unit redundant systems with gneralized repair time distributions", IEEE Trans.Rel., vol.R-17, pp.170-174, Sept.1968.

[203] M. Morrison et al., "Availability of a ν-out-of-m + r:G system", IEEE Trans.Rel., vol.R-30, pp.200-201, June 1981.

[204] T. Nakagawa, "The expected number of visits to state k before a total system failure of a complex system with repair maintenance", Oper.Res., vol.22, pp.108-116, 1974.

[205] S. Osaki, "System reliability analysis by Markov renewal processes", J.Oper.Res.Soc.Japan, vol.12, pp.127-188, May 1970.

[206] S. Osaki, "Signal-flow graphs in reliability theory", Microel. & Rel., vol.13, pp.539-541, 1974.

[207] S. Osaki et al., Reliability evaluation of some fault-tolerant computer architectures, Lecture Notes in Computer Science, no.97, Berlin/Heidelberg/New York: Springer-Verlag, 1980.

[208] S. Osaki, "Reliability evaluation of a TMR computer system with multivariate exponential failures and a general repair", Microel.& Rel., vol.22, pp.781-787, 1982.

[209] R. Ramanarayanan, "Availability of the 2-out-of-n:F system", IEEE Trans.Rel., vol.R-25, pp.43-44, Apr.1976.

[210] R. Ramanarayanan, "Reliability and availability of two general multi-unit systems", IEEE Trans.Rel., vol.R-27, pp.70-72, Apr.1978.

[211] R. Ramanarayanan et al., "n-Unit warm standby system with Erlang failure and general repair and its dual", IEEE Trans.Rel., vol.R-28, pp.173-174, June 1979.

[212] K.V. Rao et al., "Availability of an (m,N) system with repair", Microel. & Rel., vol.17, pp.571-573, 1978.

[213] S.K. Sahiar et al., "Dependability under priority repair disciplines", IEEE Trans.Rel., vol.R-25, pp.38-40, Apr.1976.

[214] M. Sasaki et al., "Reliability of intermittently used systems", IEEE Trans.Rel., vol.R-25, pp.208-209, Aug.1976.

[215] V.S. Srinivasan, "A series-parallel system", IEEE Trans.Rel., vol.R-26, pp.73-74, Apr.1977.

[216] S.S. Srivastava et al., "Stochastic behaviour of an intermittently working system with standby redundancy", Microel. & Rel., vol.10, pp.159-167, 1971.

[217] R. Subramanian et al., "Reliability of a repairable system with standby failure", Oper.Res., vol.24, pp.169-182, 1976.

[218] R. Subramanian, "Availability of a Gnedenko system", IEEE Trans.Rel., vol.R-26, pp.302-303, Oct.1977.

[219] R. Subramanian et al., "Availability of a redundant system", IEEE Trans.Rel., vol.R-27, pp.237-238, Aug.1978.

[220] R. Subramanian et al.,"On a series parallel system", Microel. & Rel., vol.20, pp.525-527,1980.

[221] R. Subramanian et al., "Interval reliability of an n-unit system with single repair facility", IEEE Trans.Rel., vol.R-30, pp.30-34, Apr.1981.

[222] R. Subramanian et al., "Multiple-unit standby redundant repairable system", IEEE Trans.Rel., vol.R-30, pp.387-388, Oct.1981.

[223] R. Subramanian et al.,"A complex two-unit parallel system", Microel. & Rel., vol.21, pp. 273-275, 1981.

[224] D.S. Taylor, "A reliability and comparative analysis of two standby system configurations", IEEE Trans.Rel., vol.R-22, pp.13-19, Apr.1973.

[225] K. Usha et al., "Two n-unit cold-standby systems with an Erlang distribution", IEEE Trans. Rel., vol.R-29, pp.434-435, Dec.1980.

[226] I.A. Ushakov, "An approximate method of calculating complex systems with renewal", Eng. Cybernetics, no.6, pp.76-83, 1980.

[227] E.J. Vanderperre, "The busy period of a repairman attaining a (n+1) unit parallel system",

Revue Française d'Automatique, Informatique et Recherche Operationnelle, 7me année no.V-2, pp.124-126, May 1973.

[228] G.K. Varma, "Stochastic behaviour of a complex system with standby redundancy", Microel. & Rel., vol.11, pp.377-390, 1972.

[229] W.E. Vesely, "A time-dependent methodology for fault tree evaluation", Nucl.Eng. and Design, vol.13, pp.337-360, 1970.

[230] Ye.M. Volovik et al., "Optimization of the servicing discipline of a system with accumulation of failures", Eng.Cybernetics, no.2, pp.84-88, 1976.

[231] G.H. Weiss, "On certain redundant systems which operates at discrete times", Technometrics, vol.4, pp.69-74, 1962.

Systems with preventive maintenance, imperfect switching, or multi failure modes

[232] K. Adachi et al., "Availability analysis of two-unit warm stand-by system with inspection time", Microel. & Rel., vol.20, pp.449-455, 1980; "Inspection policy for two-unit parallel redundant system", Microel. & Rel., vol.20, pp.603-612, 1980.

[233] M. Agarwal et al., "Analysis of a repairable redundant system with delayed replacement", Microel. & Rel., vol.21, pp.165-171, 1981.

[234] M. Alam et al., "Optimum maintenance policy for an equipment subject to deterioration and random failure", IEEE Trans.Systems, Man. and Cybernetics, vol.SMC-4, pp.172-175, March 1974.

[235] J. Ansell et al., "3-State and 5-state reliability models", IEEE Trans.Rel., vol.R-29, pp.176-177, June 1980.

[236] R. Barlow et al., "Optimum preventive maintenance policies", Op.Research, vol.8, pp.90-100, 1960.

[237] A. Birolini, "Spare parts reservation of components subjected to wear-out and/or fatigue according to a Weibull distribution", Nuclear Eng. & Design, vol.27, pp.293-298, 1974.

[238] S.M. Brodi et al., "Reliability of systems with a variable utilization mode", Eng.Cybernetics, no.5, pp.30-35, 1967.

[239] D. Chaudhuri et al., "Preventive maintenance interval for optimal reliability of deteriorating system", IEEE Trans.Rel., vol.R-26, pp.371-372, Dec.1977.

[240] W.K. Chung, "A k-out-of-N redundant system with common-cause failures", IEEE Trans.Rel., vol.R-29, p.344, Oct.1980.

[241] W.K. Chung, "An availability calculation of k-out-on-N redundant system with common-causes failures and replacement", Microel. & Rel., vol.20, pp.517-519, 1980.

[242] W.K. Chung, "A k-out-of-N:G three-state unit redundant system with common-cause failures and replacements", Microel. & Rel., vol.21, pp.589-591, 1981.

[243] W.K. Chung, "A two non-identical three-state units redundant system with common-cause failures and one standby unit", Microel. & Rel., vol.21, pp.707-709, 1981.

[244] B. Courtois, "Disponibilité de systèmes redondants périodiquement maintenus", Université de Grenoble, RR.no.148, Ensimag, 1979.

[245] A.D. Dharmadhikari et al., "Stochastic analysis of 2-unit system subjected to two types of failure", Microel. & Rel., vol.20, pp.343-345, 1980.

[246] B.S. Dhillon, "A 4-unit redundant system with common-cause failures", IEEE Trans.Rel., vol.R-26, pp.373-374, Dec.1977.

[247] B.S. Dhillon, "A system with two kinds of 3-state elements", IEEE Trans.Rel., vol.R-29, p.345, Oct.1980.

[248] Y.M.I. Dirickx et al., "Reliability of a repairable system with redundant units and preventive maintenance", IEEE Trans.Rel., vol.R-28, pp.170-171, June 1979.

[249] M.R. Dyer et al., "A note on the reliability of a system with spares which operates at discrete times", Technometrics, vol.12, pp.702-705, 1970.

[250] E.V. Dzirkal et al., "Calculation of the reliability of a redundant group with unreliable reswitching and incomplete monitorin", Eng.Cybernetics, no.6, pp.84-91, 1980.

[251] E.A. Elsayed et al., "Repairable systems with one standby unit", Microel. & Rel., vol.19, pp.243-245, 1979.

[252] R.C. Garg, "A complex system with twoo types of failure & repair", IEEE Trans.Rel., vol. R-26, pp.299-300, Oct.1977.

[253] B.V. Gnedenko et al., "The reliability of a redundant system with renewal and preventive maintenance", Eng.Cybernetics, no.1, pp.53-57, 1975.

[254] B.V. Gnedenko et al., "The duration of failure-free operation of a duplicated system with renewal and preventive maintenance", Eng.Cybernetics, no.3, pp.65-69, 1976.

[255] L.R. Goel, "Analysis of a two-unit standby system with three modes", Microel.& Rel., vol.23, pp.1029-1033, 1983.

[256] L.R. Goel et al., "A multi failure mode system with repair and replacement policy", Microel. & Rel., vol.23, pp.809-812, 1983.

[257] M.N. Gopalan et al., "Probabilistic analysis of a two-unit system with a single service facility for preventive maintenance and repair", Oper.Res., vol.23, pp.173-177, 1975.

[258] M.N. Gopalan, "Probabilistic analysis of a system with two dissimilar units subject to preventive maintenance and a single facility", Oper.Res., vol.23, pp.534-548, 1975.

[259] M.N. Gopalan et al., "Two 1-server n-unit systems with preventive maintenance and repair", IEEE Trans.Rel., vol.R-26, pp.127-128, June 1977.

[260] M.N. Gopalan et al., "Stochastic behaviour of a 2-unit system with 1-server subject to delayed maintenance", Microel. & Rel., vol.17, pp.587-589, 1978.

[261] M.N. Gopalan et al., "2-unit system with 1-repair facility subject to preventive maintenance", IEEE Trans.Rel., vol.R-27, p.77, Apr.1978.

[262] M.N. Gopalan et al., "Availability of 1-server 2-dissimilar unit system with slow switch", IEEE Trans.Rel., vol.R-27, pp.230-231, Aug.1978; "Availability analysis of 1-server n-unit system with slow switch", IEEE Trans.Rel., vol.R-27, pp.231-232, Aug.1978; "Availability analysis of 1-server n-unit with slow switch subject to maintenance", IEEE Trans.Rel., vol. R-29, p.189, June 1980.

[263] M.N. Gopalan et al., "1-server multi-component system with adjustable repair rate", IEEE Trans.Rel., vol.R-29, pp.185-186, June 1980; "2-unit system with delayed repair facility", IEEE Trans.Rel., vol.R-29, p.188, June 1980.

[264] M.N. Gopalan et al., "Busy-period analysis of a one-server two-unit system subjected to non-negligible inspection time", Microel.& Rel., vol.23, pp.453-465, 1983.

[265] A.K. Govil, "Reliability of a standby system with common-cause failure and scheduled maintenance", Microel.& Rel., vol.21, pp.269-271, 1981.

[266] P.P. Gupta, "Complex system reliability with general repair time distributions under preemptive repeat repair discipline", Microel. & Rel., vol.12, pp.145-150, 1973; "Complex system reliability with exponential repair time distributions under head-of-line-repair-discipline", Microel. & Rel., vol.12, pp.151-158, 1973.

[267] P.P. Gupta et al., "Operational availability of a complex system with two types of failure under different repair preemptions", IEEE Trans.Rel., vol.R-30, pp.484-485, Dec.1981; "Probabilistic analysis of a multi-component system with opportunistic repair", IEEE Trans. Rel., Vol.R-30, pp.487-488, Dec.1981.

[268] R.K. Gupta et al., "Complex system with preeemptive-repeat repair", IEEE Trans.Rel., vol. R-28, p.367, Dec.1979.

[269] S.M. Gupta et al., "Reliability analysis of a two-unit cold standby redundant system with two operating modes", Microel.& Rel., vol.22, pp.747-758, 1982; "Stochastic behaviour of a two-unit cold standby system with three modes and allowed down time", Microel.& Rel., vol.23, pp.333-336, 1983.

[270] S.M. Gupta et al., "Switch failure in a two-unit standby redundant system", Microel.& Rel., vol.23, pp.129-132, 1983.

[271] K. Hazeghi, Optimale Unterhaltspolitik für komplexe Systeme , IOR der ETHZ, Bern: Haupt-Verlag, 1979.

[272] C. Henin, "Double failure and other related problems in standby redundancy", IEEE Trans. Rel., vol.R-21, pp.35-40, Febr.1972.

[273] C.I. Hwang et al., "Optimal scheduled-maintenance policy based on multiple-criteria decision-making", IEEE Trans.Rel., vol.R-28, pp.394-399, Dec.1979.

[274] T. Itoi et al., "N-unit parallel redundant system with correlated failure and single repair facility" Microel. & Rel., vol.17, pp.279-285, 1978.

[275] T. Itoi et al., "Behaviour of a two correlated units redundant system with many types of failure" Microel. & Rel., vol.17, pp.517-522, 1978.

[276] A. Jain et al., "Comparison of replacement strategies for items that fail", IEEE Trans.Rel., vol.R-23, pp.247-251, Oct.1974.

[277] D.L. Jaquette et al., "Initial provisioning of a standby system with deteriorating and repairable spares", IEEE Trans.Rel., vol.R.21, pp.245-247, Nov.1972.

[278] S. Kalpakam et al., "General 2-unit redundant system with random delays", IEEE Trans.Rel., vol.R-29, pp.86-87, Apr.1980.

[279] D.V.S. Kapil et al., "Intermittently used 2-unit redundant system with PM". IEEE Trans.Rel., vol.R-29, pp.277-278, Aug.1980.

[280] K.R. Kapoor et al., "First uptime and disappointment time joint distribution of an intermittently used System", Microel. & Rel., vol.20, pp.891-893, 1980.

[281] P.K. Kapur et al., "A 2-dissimilar-unit redundant system with repair and preventive maintenance", IEEE Trans.Rel., vol.R-24, p.274, Oct.1975.

[282] P.K. Kapur et al., "Effect of standby redundancy of system reliability", IEEE Trans.Rel., vol.R-25, pp.120-121, June 1976.

[283] P.K. Kapur et al., "Joint optimum preventive-maintenance and repair-limit replacement policies, IEEE Trans.Rel., vol.R-29, pp.279-280, Aug.1980.

[284] N.K. Kashyap, "Stochastic behaviour of an intermittently working eletronic equipment with imperfect switching", Microel. & Rel., vol.12, pp.45-50, 1973.

[285] Z.S. Khalil, "On the reliability of a two-unit redundant system with random switchover time and two types of repair", Microel. & Rel., vol.16, pp.159-160, 1977.

[286] D.N. Khandelwal, "Optimal periodic maintenance policy for machines subjected to deterioration and random breakdown", IEEE Trans.Rel., vol.R-28, pp.328-330, Oct.1979.

[287] M. Kodama, "Probabilistic analysis of a multicomponent series-parallel system under preemptive repeat repair discipline", Oper.Res., vol.24, pp.500-515, 1976.

[288] J.M. Kontoleon et al., "Reliability analysis of a system subject to partial and catastrophic failures", IEEE Trans.Rel., vol.R-23, pp.277-278, Oct.1974.

[289] J.M. Kontoleon et al., "Availability of a system subjected to irregular short supervision" IEEE Trans.Rel., vol.R-24, pp.278-280, Oct.1975.

[290] V.V. Kozlov et al., "Optimal servicing of renewable systems, Part I and Part II", Eng.Cybernetics, no.3, pp.62-66 and no.4, pp.53-58, 1978.

[291] A. Kumar, "Steady-state profit in a 2-unit standby system", IEEE Trans.Rel., vol.R-25, pp.105-108, June 1976.

[292] A. Kumar, "Steady-state profit in several 1-out-of-2:G systems", IEEE Trans.Rel., vol.R-26, pp.366-369, Dec.1977.

[293] A. Kumar et al., "Analysis of a 2-unit standby redundant system with two types of failures", IEEE Trans.Rel., vol.R-27, pp.301-302, Oct.1978.

[294] A. Kumar et al., "Analysis of a two-unit maintained series system with imperfect detection on failures", Microel. & Rel., vol.19, pp.329-331, 1979.

[295] A. Kumar et al., "Behaviour of a two-unit standby redundant system with imperfect switching and delayed repair", Microel. & Rel., vol.20, pp.315-321, 1980.

[296] A. Kumar et al., "Availability of a two-unit standby system with switchover time and proper initialization of connect switching", Microel. & Rel., vol.21, pp.113-115, 1981.

[297] A. Kumar et al., "Stochastic behaviour of a two-unit redundant system with switchover time", Microel. & Rel., vol.21, pp.717-725, 1981.

[298] A. Kumar, "Availability of a complex system under several repair preemptions", IEEE Trans. Rel. vol.R-30, pp.485-486, Dec.1981.

[299] Laprie et al., "Parametric analysis of 2-unit redundant computer systems with corrective and preventive maintenance", IEEE Trans.Rel., vol.R-30, pp.139-144, June 1981.

[300] M.I. Mahmoud et al., "The effect of preventive maintenance to a standby system with two types of failures", Microel. & Rel., vol.23, pp.149-152, 1983.

[301] M.I. Mahmoud et al., "Stochastic behaviour of a 2-unit standby redundant system with imperfect switchover and preventive maintenance", Microel. & Rel., vol.23, pp.153-156, 1983.

[302] Y.K. Malaiya, "Linearly correlated intermittent failures", IEEE Trans.Rel., vol.R-31, pp. 211-215, June 1982.

[303] C. Maruthachalam et al., "Steady-state availability of a system with two subsystems working alternately", Microel. & Rel., vol.2, pp.935-940,1982.

[304] M. Mazumdar, "Reliability of two-unit redundant repairable systems when failures are revealed by inspections", SIAM J.Appl.Math., vol.19, pp.637-647, Dec.1970.

[305] J.B. de Mercado, "Reliability prediction studies of complex systems having many failed states", IEEE Trans.Rel., vol.R-20, pp.223-230, Nov.1971.

[306] M. Messinger et al., "Techniques for optimum spares allocation: a tutorial review", IEEE Trans.Rel., vol.R-19, pp.156-166, Nov.1970.

[307] H. Mine et al., "An optimal maintenance policy for a 2-unit parallel system with degraded states", IEEE Trans.Rel., vol.R-23, pp.81-86, June 1974.

[308] H. Mine et al., "Stochastic behaviour of two-unit redundant systems which operate at discrete times", Microel. & Rel., vol.15, pp.551-554, 1976.

[309] H. Mine et al., "Interval reliability and optimum preventive maintenance policy", IEEE Trans. Rel., vol.R-26, pp.131-133, June 1977.

[310] H. Mine et al., "Preventive replacement of an intermittently-used system", IEEE Trans.Rel., vol.R-30, pp.391-392, Oct.1981.

[311] K. Murari et al., "A system with two subsystems working in alternating periods", Microel. & Rel., pp.405-412, 1982.

[312] T. Nakagawa et al., "Optimum preventive maintenance policies for a 2-unit redundant system", IEEE Trans.Rel., vol.R-23, pp.86-91, June 1974.

[313] T. Nakagawa et al., "Optimum preventive maintenance policies maximizing the mean time to the first system failure for a two-unit standby redundant system", J. Optimization theory and Appl., vol.14, pp.115-129, July 1974.

[314] T. Nakagawa et al., "Stochastic behaviour of a two-dissimilar-unit standby redundant system with repair maintenance", Microel. & Rel., vol.13, pp.143-148, 1974.

[315] T. Nakagawa et al., "Stochastic behaviour of a two-unit standby redundant system with imperfect switchover", IEEE Trans.Rel., vol.R-24, pp.143-146, June 1975.

[316] T. Nakagawa et al., "Analysis of a repairable system which operates at discrete times", IEEE Trans.Rel., vol.R-25, pp.110-112, June 1976.

[317] T. Nakagawa et al., "Reliability analysis of intermittently used systems when failures are detected only during a usage period", Microel. & Rel., vol.15, pp.35-38, 1976.

[318] T. Nakagawa, "A 2-unit repairable redundant system with switching failure", IEEE Trans. Rel., vol.R-26, pp.128-130, June 1977.

[319] T. Nakagawa, "Optimum preventive maintenance policies for repairable systems", IEEE Trans. Rel., vol.R-26, pp.168-173, Aug.1977.

[320] T. Nakagawa, "Reliability analysis of standby repairable systems when an emergency occurs", Microel. & Rel., vol.17, pp.461-464, 1978.

[321] T. Nakagawa, "Optimum policies when preventive maintenance is imperfect", IEEE Trans.Rel., vol.R-28, pp.331-332, Oct.1979.

[322] T. Nakagawa, "Imperfect preventive-maintenance", IEEE Trans.Rel., vol.R-28, p.402, Dec. 1979.

[323] T. Nakagawa, "Mean time to failure with preventive maintenance", IEEE Trans.Rel., vol.R-29, p.341, Oct.1980.

[324] T. Nakagawa, "Replacement policies for a unit with random and wearout failures", IEEE Trans.Rel., vol.R-29, pp.342-344, Oct.1980.

[325] T. Nakagawa, "Replacement models with inspection and preventive maintenance", Microel. & Rel., vol.20, pp.427-433, 1980.

[326] D.G. Nguyen et al., "Optimal maintenance policy with imperfect preventive maintenance", IEEE Trans.Rel., vol.R-30, pp.496-497, Dec.1981.

[327] D.G. Nguyen et al., "Optimal preventive maintenance policies for repairable systems", Oper. Res., vol.29, pp.1181-1194, 1981.

[328] D.S. Nielsen et al., "Unreliability of a standby system with repair and imperfect switching", IEEE Trans.Rel., vol.R-23, pp.17-24, Apr.1974.

[329] A. Omar et al., "Nonloaded duplexing taking switching time into account", Eng.Cybernetics, no.4, pp.310-313, 1966.

[330] S. Osaki et al., "A two-unit standby redundancy system with repair and preventive maintenance", J.Appl.Prob., vol.7, pp.641-648, Dec.1970.

[331] S. Osaki, "On a two-unit standby-redundant system with imperfect switchover", IEEE Trans. Rel., vol.R-21, pp.20-24, Febr.1972.

[332] S. Osaki, "Reliability analysis of a two-unit standby-redundant system with preventive maintenance", IEEE Trans.Rel., vol.R-21, pp.24-29, Febr.1972.

[333] S. Osaki, "An intermittently used system with preventive maintenance", J.Oper.Res.Soc. Japan, vol.15, pp.102-111, June 1972.

[334] S. Osaki et al., "Repair limit suspension policies for a two-unit standby redundant system with two phase repairs", Microel. & Rel., vol.16, pp.41-45, 1977.

[335] W.P. Pierskalla et al., "A survey of maintenance models: the control and surveillance of deteriorating systems", Naval Res.Logistic Q., vol.23, pp.353-388, 1976.

[336] S. Prakash, "Stochastic behaviour of a redundant electronic equipment with imperfect switching and opportunistic repairs" and "Some reliability characteristic of a standby redundant equipment with imperfect switching", Microel.& Rel., vol.9, pp.413-418 and pp.419-423, 1970.

[337] C.L. Proctor et al., "A repairable 3-state device", IEEE Trans.Rel., vol.R-25, pp.210-211, Aug.1976.

[338] C.L. Proctor et al., "The analysis of a four-state system", Microel. & Rel., vol.15, pp.53-55, 1976.

[339] D.V. Rozhdestvenskiy et al., "Reliability of a duplicated system with renewal and preventive maintenance", Eng.Cybernetics, no.8, pp.475-479, 1970.

[340] V.V.S. Sarma et al., "Optimal maintenance policies for machines subject to deterioration and intermittent breakdowns", IEEE Trans.Systems, Man. and Cybernetics", vol.SMC-5, pp. 396-398, May 1975.

[341] W.G. Schneeweiss, "On the mean duration of hidden faults in periodically checked systems" IEEE Trans.Rel., vol.R-25, pp.346-348, Dec.1976.

[342] W.G. Schneeweiss, "Duration of hidden faults in randomly checked systems", IEEE Trans. Rel., vol.R-26, pp.328-330, Dec.1977.

[343] W.G. Schneeweiss, "Reliability models for switches for duplicated computer modules", Microel. & Rel., vol.20, pp.571-579, 1980.

[344] C.M. Shama, "Time dependent solution of a complex standby redundant system under preemptive repeat repair discipline", IEEE Trans.Rel., vol.R-23, pp.283-285, Oct.1974.

[345] Y.S. Sherif et al., "Optimal maintenance models for systems subject to failure - a survey", Naval Res.Logistics Q., vol.28, pp.47-74, 1981; "Optimal maintenance schedules of systems subject to stochastic failure", Microel.&Rel., vol.22, pp.15-29, 1982.

[346] C. Singh et al., "Reliability modelling in systems with non-exponential down time distributions", IEEE Trans.Power App.Syst., vol.PAS-92, pp.790-800, 1973.

[347] C. Singh et al., "The method of stages for non-Markov models", IEEE Trans.Rel., vol.R-26, pp.135-137, June 1977.

[348] I.P. Singh et al., "3-Component intermittent system", IEEE Trans.Rel., vol.R-28, pp.415-416, Dec.1979.

[349] J. Singh, "Effect of switch failure on 2 redundant systems", IEEE Trans.Rel., vol.R-29, pp.82-83, Apr.1980.

[350] S.P. Singh et al., "Two-unit redundant system with random switchover time and two types of repair", Microel. & Rel., vol.19, pp.325-328, 1979.

[351] S.M. Sinha et al., "2-Unit redundant system with delayed switchover and two types of repairs, IEEE Trans.Rel., vol.R-28, p.417, Dec.1979.

[352] S.M. Sinha et al., "Optimum preventive maintenance policies for 2-unit redundant system with repair and post-repair", Microel. & Rel., vol.20, pp.887-890, 1980.

[353] V.S. Srinivasan "A standby redundant model with noninstantaneous switchover", IEEE Trans. Rel., vol.R-17, pp.175-178, Sept.1968.

[354] V.S. Srinivasan,"A cold-standby redundant system with delayed switchover and preventive maintenance, IEEE Trans.Rel., vol.R-26, pp.238-239, Aug.1977.

[355] A. Streller, "Stationary interval reliability of a redundant system with renewal and preventive maintenance, Eng.Cybernetics, no.3, pp.72-77, 1979.

[356] R. Subramanian et al.,"Reliability of a 2-unit standby redundant system with repair maintenance and standby failure", IEEE Trans.Rel., vol.R-24, pp.139-142, June 1975.

[357] R. Subramanian et al., "Availability of a 2-unit redundant system with preventive maintenance", IEEE Trans.Rel., vol.R-27, pp.73-74, Apr.1978.

[358] R. Subramanian, "Availability of 2-unit system with preventive maintenance and one repair facility", IEEE Trans.Rel., vol.R-27, pp.171-172, June 1978.

[359] R. Subramanian et al., "On a two unit standby redundant system with imperfect switchover", Microel. & Rel., vol.17, pp.585-586, 1978.

[360] R. Subramanian et al., "A 2-unit priority redundant system with preemptive resume repair", IEEE Trans.Rel., vol.R-29, pp.183-184, June 1980.

[361] R. Subramanian et al., "Complex two-unit system with preventive maintenance", Microel. & Rel., vol.21, pp.559-567, 1981.

[362] R. Subramanian et al., "A redundant system with non-instantaneous switchover and preparation time for the repair facility", Microel. & Rel., vol.21, pp.593-596, 1981.

[363] R. Subramanian et al.,"An n-unit standby redundant system with r repair facilities and preventive maintenance", Microel. & Rel., vol.22, pp.367-377, 1982.

[364] R. Subramanian et al., "Stochastic behaviour of a two-unit repairable system subject to two types of failure and inspection", "Analysis of a two-unit repairable system with random inspection subjected to two types of failures", Microel.& Rel., vol.23, pp.445-447 and 449-451, 1983.

[365] Y. Sugasaw, "Light maintenance for a two-unit parallel redundant system with bivariate exponential lifetimes", Microel. & Rel., vol.21, pp.661-670, 1981.

[366] I.M. Titenko, "The reliability of a periodically monitored system with preventive maintenance", Eng.Cybernetics, no.2, pp.113-119, 1981.

[367] Yu.D. Umrikhin, "Reliability of a periodically controlled system", Eng.Cybernetics, no.6, pp.60-67, 1978.

[368] B.Ya. Vtorova-Karevskaya e al., "On the reliability of systems admitting failures of two kinds", Eng.Cybernetics, no.4, pp.133-135, 1979.

[369] M. Yamashiro, "A repairable multistate device with general repair time", "A multistate system with general repair time distribution", IEEE Trans.Rel., vol.R-29, p.276, Aug. 1980 and p.453, Dec.1980; "Two repairable multistate devices with general repair-time distributions", IEEE Trans.Rel., vol.R-30, p.204, June 1981.

[370] M. Yamashiro, "Analysis of a degraded multistate system with general repair-time distributions", "A multistate system with several failure modes and cold standby units", Microel.& Rel., vol.20, pp.647-650 and 673-677, 1980.

[371] W.T. Yang, "A Reliability model for dependent failures in parallel redundant systems", IEEE Trans.Rel., vol.R-23, pp.286-287, Oct.1974.

[372] Y. Yonehara et al., "Reliability analysis of a 2-out-of-n: F system with repairable primary and degradation units", Microel.& Rel., vol.22, pp.1081-1097, 1982.

INDEX